Geometry of Grief

Geometry of Grief

REFLECTIONS ON MATHEMATICS, LOSS, AND LIFE

Michael Frame

The University of Chicago Press
Chicago and London

The University of Chicago Press, Chicago 60637
The University of Chicago Press, Ltd., London

Published 2021
Paperback edition 2023
Printed in the United States of America

32 31 30 29 28 27 26 25 24 23   1 2 3 4 5

ISBN-13: 978-0-226-80092-9 (cloth)
ISBN-13: 978-0-226-82648-6 (paper)
ISBN-13: 978-0-226-80108-7 (e-book)
DOI: https://doi.org/10.7208/chicago/9780226801087.001.0001

Library of Congress Cataloging-in-Publication Data

Names: Frame, Michael, author.
Title: Geometry of grief : reflections on mathematics, loss, and life / Michael Frame.
Description: Chicago : University of Chicago Press, 2021. | Includes bibliographical references and index.
Identifiers: LCCN 2021007566 | ISBN 9780226800929 (cloth) | ISBN 9780226801087 (ebook)
Subjects: LCSH: Fractals. | Grief. | Geometry. | Mathematics—Social aspects.
Classification: LCC QA614.86 .F796 2021 | DDC 514/.742—dc23
LC record available at https://lccn.loc.gov/2021007566

♾ This paper meets the requirements of ANSI/NISO Z39.48-1992 (Permanence of Paper).

Contents

Prologue

Dad, that's really scary.

"Look for the very brightest one in the sky."

"Beside the tree, about halfway up? Is that it, Ruthie?"

"That's it. That's Venus. It's a planet, a whole world, almost as big as the earth. And it's cloudy all the time. No one has ever seen land on Venus."

"If it's cloudy all the time, Venus must be cold."

"Not necessarily. Venus is closer to the sun than Earth is. Maybe the clouds hold in the heat and it's very hot there."

"Oh, I see. The sky is clear tonight, so we'll get cooler than we would if it was cloudy."

"That's right, Mikey. Do you want to go inside now?"

"Are there other planets in the sky?"

"Not tonight."

"Can we stay outside and watch lightning bugs?"

"Sure."

This was an evening late in the summer of 1958. The sky, purple deepening to indigo, showed a few pinpoint stars and a much brighter dot, Venus. We'd had dinner with my grandmother and my Aunt Ruthie, my dad's sister, at their house in South Charleston, West Virginia. I was seven, my sister Linda was four, our brother Steve was two. Only Ruthie and I were in the backyard. The others were on the front porch, "visiting,"

Mom called it. We lived in St. Albans, West Virginia, only about eight miles away, and saw my grandmother and Ruthie often. Just why the adults visited wasn't clear to me. What could they talk about? They just gossiped about their neighbors and other family members.

Ruthie and I were different. That afternoon we'd sat in the kitchen garden, and the purposeful march of ants and the random jumps of grasshoppers entranced us. I constructed elaborate natural histories to explain their behaviors; Ruthie proposed much simpler alternatives. She never used the term "Occam's razor," but she had begun to teach me the beauty of simple explanations. And also the likelihood of economy: a Rube Goldberg machine—a complicated contraption that takes up a whole room and performs a simple task like cracking an egg—has many points of potential failure. My complex pathways were good mental exercises, maybe, but did I really think nature would be that silly? Years later, I understood that Ruthie had started me on the path to becoming a scientist. She thought curiosity is the most important trait of the mind; that the curiosity of a child, the twists and turns of young logic when the child unpacks aspect and dynamic of the wide world, is the most beautiful thing an adult can see. Mom and Dad, grandparents, other aunts and uncles, encouraged curiosity, but Ruthie cultivated it, mixed in some skepticism, and always found a book for me to read about the topic of current interest. Ruthie set me on the way that, sixty years later, has led me to write this story.

In elementary school career discussions, against my classmates' police officer, firefighter, and park ranger (astronaut wasn't a career then—yes, I'm ancient), I offered physicist or mathematician or astronomer. But really, at that age every kid is a naturalist. A summer morning in neighboring woods revealed wonders without end. The optimism of childhood knew no bounds. My parents' finances, though limited, afforded opportunity for creative explorations. To measure the output of a thermocouple (a copper wire and a steel wire twisted together that convert heat into a weak electrical current), the father of another student bought an expensive multimeter. I made a galvanometer: two magnetized needles stuck through a small cardboard rectangle suspended by thread in a coil of wire. Who had more fun detecting the tiny current?

Ruthie didn't help me design the experiments—Dad did that, and let me set up a small lab in a corner of his workshop—but Ruthie helped me realize that I *could* do experiments and answer some of my own questions.

Late in my eleventh year, Ruthie got sick. Hodgkin's lymphoma, survivable now but not so much in the early 1960s. She was treated, with the chemotherapy drug Mustargen, I believe, but lived only a few more months in some misery and died early in my twelfth year. I visited Ruthie when she was sick, but I couldn't do much. I stood beside her bed, rested my little hand on her forearm and tried to talk with her. But I couldn't think of anything to say. At home after these visits, Mom hugged me, stroked my hair. I knew I should have talked more with Ruthie. She had done so much for me, and she needed me now. She needed me to talk with her because I was her favorite. Later I understood that Mom was working through her own grief. She knew the situation far better than I did, knew this disease would win and Ruthie would lose. Dad began to talk with me about his sister's illness. He was straightforward: Ruthie was going to die. I appreciated his honesty. No nonsense about Ruthie going away, or—worse—going to live with the angels.[1] Her life would end, and soon. "This isn't fair. There's so much more for Ruthie and me to do. She promised we'd get a telescope to look at the planets. I've saved my allowance for six months already. This just isn't fair."

"Son, life isn't fair. Ruthie isn't sick because she did anything bad. She just got sick. Sometimes good things happen, sometimes bad things happen. All we can do is try to make a few more good things happen and a few less bad things happen. But a lot of things that happen to us, we can't do anything about."

"Dad, that's really scary."

"Yes, son, it really is."

That night I thought of a plan. I'd work very, very hard. Study all the time, no more hide-and-seek or silly stories told to little kids. I'd finish high school years early, go to college, then graduate school and medical school, become a medical researcher, find a cure for Hodgkin's lymphoma, administer it to Ruthie, and save her. In one version of the fantasy, I flew in a helicopter from my university laboratory to Ruthie's hospital. I was so pleased with my plan. I told Mom and said I'd tell Ruthie not to worry, that I'd save her. I expected Mom to be happy, but she looked very sad, told me I couldn't tell Ruthie.

"Why not? Don't you want her to know she'll be alright?"

"Mikey, I don't want you to get her hopes up." A lie, but a gentle, sweet lie. "No matter how hard you work, you might not be able to save Ruthie."

Logically I knew Mom was right. I'd gone to the library in Charleston, found an oncology book (I'd asked Mom the scientific name for the study of cancer), and found the Hodgkin's survival statistics. They weren't encouraging. But I wasn't able to imagine a world without Ruthie. We had years of exploration still to do. And besides, how could Ruthie leave her sweet mother, Luverna Frame, the kindest, gentlest adult in my world? There had to be a way out of this, and I would find it.

But Ruthie died. Dad was at the hospital with her, holding her hand, when she died. When he came home, his expression told me all I needed to know. He told Mom, Linda, and Steve. They cried; I didn't. Eventually Mom said that Ruthie had been terribly sick, would never be well again, so it was better that she wasn't suffering anymore. "Ruthie was suffering?" Linda wailed.

Then she and Steve began to run around and shriek. Eventually they calmed down to sobs. But I'd known that Ruthie was miserable. Waiting at the hospital hall outside her room while Dad checked that it was okay for me to come in, sometimes I heard her moan. She'd suffered, and now she didn't. Was the peace of non-existence better than pain with little relief? Big puzzles for a twelve-year-old. Big puzzles still. . . .

Dad didn't want us kids to go to the funeral. Mom and Dad went while we stayed with Mom's parents, Burl and Lydia Arrowood. I found a bag of balloons in Gramp's workshop. Gramp was a jeweler and repaired clocks and watches. Because he used a gas torch to melt some alloys, his workshop had a gas jet. I filled a balloon with gas, tied it off, walked into the front yard away from the trees and let the balloon fly. The symbolic content of this was melancholy: it represented all the experiments that Ruthie and I had planned to do, that now were lost forever. It represented the closing of a door.

So I closed myself off from the world. I could no longer help Ruthie, but maybe I could help other people. All I did was read and study science. Mom and Dad tried to get me to run around outside. They said Linda and Steve missed me, but I don't think they did. All summer long they were outside, awakened by an early serenade of blue jays and catbirds, games of tag and hide-and-seek interrupted by dusk with its drifting fireflies. No, they didn't need me.

And now I had a goal: I could no longer help Ruthie, but I could find cures for diseases and save other people. The determination of a serious twelve-year-old can be fierce, and I was fierce squared.

Later that year, I read a supplementary problem in my algebra text. For much of the weekend I tried all sorts of tricks. Eventually I found a solution, but it was clunky, mechanical, inelegant. It worked, but I knew it wasn't the solution the author intended.

After math class on Monday, I asked my teacher. She smiled, said she was happy I tried the problem, then wrote the simple, beautiful solution.

At that moment, my world folded in on itself, disappeared, and I knew what I thought was a different flavor of grief. The solution used only tricks I knew but applied them in a clever way that hadn't occurred to me. At that moment, I began to suspect I was not bright enough to be a good scientist. Determination and hard work would get me into the tribe of scientists, but would life as a supporting character be enough? Choosing that path carried the real risk that from the end of life, where I am now, a backward look would reveal decades of steady work punctuated by very few moments of modest insight. To be sure, those moments have been amazing. The pleasure of understanding some bits of the architecture of ideas is ample reward. But I wanted to do so much more.

Has my life been so different from the lives of others? For some people, aptitude and interest align and a satisfying life unfolds, enviably free from regret or second-guessing. But many of us are haunted by thoughts of a path not taken. Some choices lead us along paths that we cannot reverse. Even if we change course now, what remains of our lives will not unfold as if we had made the other choice years earlier. What might have been is beyond our reach, and we grieve this loss.

For me, the path I choose—exploring some structures of math—has afforded new perspectives on grief. I believe grieving exhibits some similarities to doing math; we'll find echoes of each in the other. Wrestling with mathematical questions has helped me to parse my own episodes of grief. That is my subject here.

In *The Doubter's Almanac*, Ethan Canin writes:

> Is the sorrow of death the same as the sorrow of knowing the pain in a child's future? What about the melancholy of music?

> Is it the same as the melancholy of a summer dusk? . . . Both we call grief. . . .
>
> But how to solve the grief I felt for my father in those last days? We think that our sorrow, like the planes we know in this world, has borders. But does it?[2]

Because geometry is, for me, the most beautiful part of math, and the part I know best, I'll focus on geometry: the geometry of grief. This is as distinct from the grief of geometry—the longing to escape a late afternoon class where the teacher plodded through a proof of "side-angle-side" in two-column format on the chalkboard—as the lyric "If it weren't for bad luck, I'd have no luck at all" is from Puccini's "Nessun dorma." In this book we'll explore some ways in which grief informs geometry and geometry informs grief.

The architecture of this project was largely in place before I looked into what others had written. A notion that's repeated often in this book is that an idea can't be unseen. Taking in others' ideas before thinking through my own experiences with grief might have limited how I understood those experiences. Only after I'd sketched out a rough draft did I read background studies on grief. Particularly useful were the evolutionary perspectives in psychologist John Archer's *The Nature of Grief*, anthropologist Barbara King's *How Animals Grieve*, and physician and evolutionary biologist Randolph Nesse's "An Evolutionary Framework for Understanding Grief."[3] Some of my ideas are similar to established notions; others differ, occasionally significantly. I'll point those out.

Am I being egotistical in placing my own thinking before that of scholars who have studied grief for years? You may disagree, but my answer is No. In the dark hours between midnight and dawn, we are alone with our thoughts. This is when we best sift through our personal grief. The first step is to understand our own experience, then see how it fits into established work.

Before anything I write can make sense, you must peer into your own grief.

* * *

For all the admiration I have for Archer, King, Nesse, and other clever scholars, I think that literature, film, and music provide a more immediate glimpse of the internal world of grief. Others have had this idea, too. In writing *The Foundations of Character*, the first systematic study of the psychology of grief, Alexander Shand had little experimental data so instead drew on poetry and literature, works by authors who were careful, thoughtful observers of human nature.[4] Archer recognized the power of some literature to give a clear expression of views with emotional weight, and he studied grief viewed through art.[5]

Stories give the most direct, nuanced, and expansive picture. I learned more about Sartre's views on existentialism from his trilogy *The Roads to Freedom* than I did from his dense philosophical tome *Being and Nothingness*.[6] I'll tell a lot of stories in the chapters that follow.

To understand how art can give a visceral sense of the depths of love and of grief, think of the lyrics of "My Skin" by Natalie Merchant, or the catch in her voice in "Beloved Wife." Think of the sad yet hopeful lyrics of "Dante's Prayer" by Loreena McKennitt. Think of the breathtaking end of "Knee 5," from *Einstein on the Beach* by Philip Glass. Some other movements of Glass's opera are more interesting musically, but the multilayered voices and matter-of-fact reading of this passage take my breath away. Music can speak directly to us about deep emotions.[7]

If you have seen Ang Lee's beautiful film *Crouching Tiger, Hidden Dragon*, think of Li Mu Bai dying in the arms of Yu Shu Lien. Li's last words, spoken to Shu Lien, "I've already wasted my whole life. I want to tell you with my last breath . . . I have always loved you. I would rather be a ghost, drifting by your side . . . as a condemned soul . . . than enter heaven without you. Because of your love . . . I will never be a lonely spirit."

Or think of the end of the film. Jen Yu is in the temple on Wudan Mountain. She stands with the bandit Lo, her lover, on a bridge above a bank of clouds. Jen asks, "Do you remember the legend of the young man?" Earlier Lo had told her:

> We have a legend. Anyone who dares to jump from the mountain, God will grant his wish. Long ago, a young man's parents were ill, so he jumped. He didn't die. He wasn't even hurt. He floated away, never to return. He knew his wish had come true. If you believe, it will happen. The elders say, "A faithful heart makes wishes come true."

Lo: "A faithful heart makes wishes come true." Jen: "Make a wish, Lo." Lo: "To be back in the desert, together again." Jen jumps from the bridge and disappears into the clouds. Yo-Yo Ma's cello sings in the "Farewell" track as Jen sails through the clouds. By now you must know that this is *not* a martial arts film. It is a story of love, loss, and grief.[8]

Or perhaps you've seen *Six Feet Under*, a five-year series about the lives of a family who run a Los Angeles funeral home. Now, you might criticize this choice as low-hanging fruit: people who run a funeral home encounter grief every workday. But each episode explores death and grief from a particular philosophical or psychological point of view. Right now I am particularly interested in the series finale, accompanied by Sia's "Breathe Me."[9] We see the unfolding and conclusion of the lives of the main characters, arcs of lives revealed, grief on many levels, grief as a reflection of love. And I must mention the adaptation, funny until it *really* wasn't, in "Flanders' Ladder," season 29, episode 21, of *The Simpsons*.

Think of the death of Yevgény Vasílevich Bazárov in Ivan Turgenev's *Fathers and Sons*, of the incandescent grief his old parents felt at his graveside.[10] Yevgény's death could have been avoided. A moment's slip and, in the world of the story, his death was certain, irreversible. Describing a small village graveyard, Turgenev writes, at the end of the novel:

> Often from the little near-by village two frail old people, a husband and wife, make their way there. Supporting each other, they walk with heavy steps; they go up to the iron railing, fall on their knees and weep long and bitterly, and long and yearningly they gaze at the silent stone beneath which their son is lying.

This picture of the parents' grief is chilling. But in the context of the larger story, we understand their heartbreak more fully. Sometimes I think the juxtaposition of this crushing emotion and Turgenev's straightforward prose reveals a kernel of great beauty buried deep in grief.

Stories can't really tell us how another person feels, but they can help us imagine how we would feel were we in their situation. This is, I believe, the basis of empathy. This is how we'll try to understand a bit about grief.

Many of the ideas we'll discuss are the results of introspection, my own experiences with grief and geometry. I'll present them as stories, mostly, rather than as abstract arguments, because I think stories are more effective ways to communicate ideas that have emotional importance. Abstract arguments can give some background, but stories make the points immediate, urgent.

Perhaps my experiences will remind you of yours. Or maybe your experiences are substantially different. Must different experiences lead to different understandings of grief? I don't know. The physical world has room enough for many vistas. How many more may inhabit our minds?

* * *

In this book we'll uncover a few ideas, most presented several times, in different contexts, and all illustrated with stories. Here's a short description of the main points we'll explore.

Grief is a response to an irreversible loss. A corollary: there is no anticipatory grief.[11] To generate grief rather than sadness, the thing lost must carry great emotional weight, and it must pull back the veil that covers a transcendent aspect of the world. Breathe out to push the fog away from a brilliant pinpoint of light. We'll focus on these three aspects of grief: it is irreversible, it carries emotional weight, and it is transcendent. Grief is not the only experience that exhibits these properties. I have no children myself, but becoming a parent, I imagine, is similarly profound: irreversible, emotionally weighty, and transcendent. The obvious additional aspect of grief is that it is a response to loss.

Grief has an evolutionary basis. We'll explore some arguments for this basis and some evidence that animals feel grief. We'll see also that literature and music offer effective, sometimes profound, windows into these experiences. This, too, marks a path to owning our grief.

The initial "aha" moment, when first we understand something, can occur only once. If the thing understood is important to us and hints at deeper mysteries, we can grieve the loss of this moment, which, once we experience it, is lost forever. Beauty seen in a mirror reflects grief. This is the key to the connection I'll make between geometry and grief.

The view of life trajectories in story space, introduced in chapter 4, gives a way to project grief and perhaps find some relief from its sting. Story space is the primary tool we develop, so here I'll mention the main points:

- All moments of our lives are immensely rich, with many—perhaps infinitely many—variables we could notice.

- We can view our lives as trajectories, parameterized by time, through story space.
- We can never simultaneously view all of the possible variables; rather, we focus on a few variables at a time, restricting our attention to a low-dimensional subspace of story space.
- Our trajectories through these subspaces are the stories we tell ourselves about our lives; they are how we make sense of our lives, but always they miss some elements of our experiences.
- Irreversible loss appears as a discontinuity, a jump, in our path through story space.
- By focusing on certain subspaces, by projecting our trajectories into these spaces, we can reduce the apparent magnitude of the jumps, and consequently find a way to confront the emotional loss and perhaps reduce its impact. We'll illustrate this last point with an example or two.

Moreover, grief is self-similar: the grief of losing a parent contains many "smaller" griefs. No more conversations, no more comparing memories of times good or bad, no more quiet walks together. Each of these is a scaled version of the response to the loss of a parent, a smaller copy that can act as a laboratory to find effective projections. Projecting outward, grief can point to actions that can help others. My most optimistic thought is that some of the energy of grief can be redirected in this way. Small steps or large steps, but steps all the same.

This book is a love song to my late parents, to friends we've lost and to cats we've lost. And the book is a love song to geometry, the brightest point in my mind. In old age, my understanding of geometry disappears more with each passing year and adds complex fractures to my breaking heart.

The bits of geometry that I present here are not a simple

recipe to help you get through grief, but they are a sketch of a perspective that has helped me. Perhaps they are signposts of ways you can adapt this approach to blunt your own suffering. And maybe they will help you see geometry in parts of your life where before you saw none.

1
Geometry

I miss the way I saw trees before.

Suppose the season is early spring, the time is dusk, and you are in a park you do not know well. What do you see when you look up from this page? A complicated pattern of lighter and darker shapes resolves, perhaps, into rough cylinders that are tree trunks; large branches, smaller branches, twigs; raggedy bits of planes that are leaves. Then flowers and grasses. The recognition of geometries helps us identify, or at least give names to, our surroundings.

We see changes in apparent shapes and recognize dynamics—leaves and branches, for example, dancing on a gentle breeze.

At the top of a tall tree, leaves still are illuminated, even though the trunk is in shadow. We speak of darkness falling but see here how dusk rises (and, if we return in the morning, how dawn falls). The geometry of the sun and earth reveal simple things we may have missed about the world.

For centuries artists have had good intuition about geometries. Here are some examples. Spend a few minutes with Google and you can find many, many more.

Built in the ninth century and rebuilt in the thirteenth, the Alhambra in Granada, Spain, is a beautiful expression of Islamic art and architecture. Many of the decorative tilings, including the one sketched on the next page, are tessellations of the plane.

These are shapes that could fill up the whole plane without overlaps or gaps, each shape meeting others only along (parts of) their edges. The squares of a chessboard and hexagons of a honeycomb are the most familiar, but there are others.

Branko Grünbaum and Geoffrey Shephard's *Tilings and Patterns* (at seven hundred pages this book earns the descriptor "encyclopedic") provides a wealth of examples, some in art but most from math.[1] Altogether there are seventeen distinct patterns, now given the evocative name "wallpaper groups." That only seventeen patterns exist was proved at the end of the nineteenth century, but Muslim artists were familiar with these tilings hundreds of years before Russian crystallographer and mathematician Evgraf Fedorov wrote his proof.[2] Sometimes artists develop good intuitions long before mathematicians prove their insights are correct.

Another example of how geometry and art interact is based on similar triangles. From geometry class recall that two triangles are similar if they have the same shape, even if they are of different sizes. A shape is called *self-similar* if it is made of pieces each of which is similar to the whole shape. The picture above, on the left, of triangles within triangles is the *Sierpinski gasket*, one of the best-known self-similar shapes. To see its self-similarity, note that the gasket consists of three pieces—lower left, lower right, and upper center—each similar to the whole gasket. We'll say more about the gasket in chapter 3.

Fractals, a class of shapes first recognized by mathematician Benoit Mandelbrot, are shapes made up of pieces that are each similar in some way to the whole. A bit of a coastline viewed from up close looks like a larger expanse of coastline viewed from far away; a frond of a fern looks like a small fern; the two-meter-long strand of your DNA folds up into a cell's nucleus, with a diameter about a millionth of the strand's length, by executing the same pattern of folds over and over at smaller and smaller sizes. These are fractals found in nature. The simplest fractals are the self-similar shapes, such as the Sierpinski gasket.

The circular sketch next to the gasket is of a thirteenth-century tiling in an Italian cathedral, with six shapes that approximate curved Sierpinski gaskets surrounded by a ring of many smaller gaskets.[3] (To make this sketch, I measured and

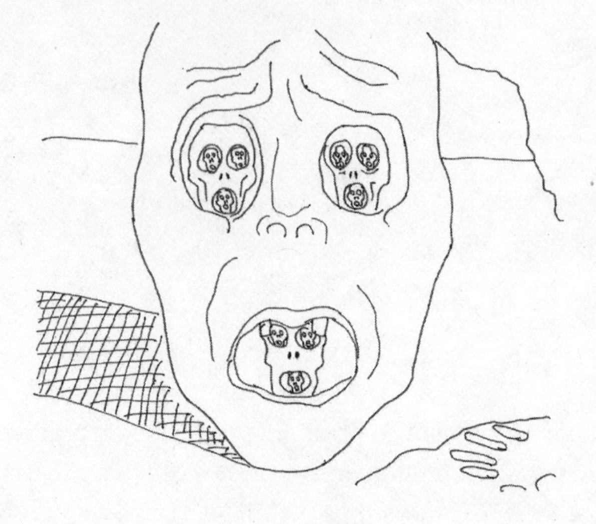

drew the main features, then filled in the rest by eye. This took some time. But the original was carved by hand, piece by piece, and then fitted together. When I remember this, the hour or so I spent on the sketch does not seem so bad.)

Artists have thought of self-similarity for many centuries. Why? Because much of nature exhibits self-similarity, and artists are careful observers of nature.

A more recent example of self-similarity is Salvador Dalí's 1940 painting *The Visage of War*, a depiction of the infinite horrors of the Spanish Civil War. In Dalí's painting, we see a face whose eye sockets and mouth contain faces, whose eye sockets and mouths contain faces, and so on for several more levels. The pattern is much like that of the Sierpinski gasket—a recurring triangle of features at, in this case, upper left, upper right, and lower center. Dalí's painting is much scarier than this sketch, with a tangle of snakes coming around both sides of the bodyless head.[4]

In a preliminary study for the painting, only the mouth contains another face. One eye socket contains the rings of a tree trunk, the other a honeycomb. Dalí discovered that the repeti-

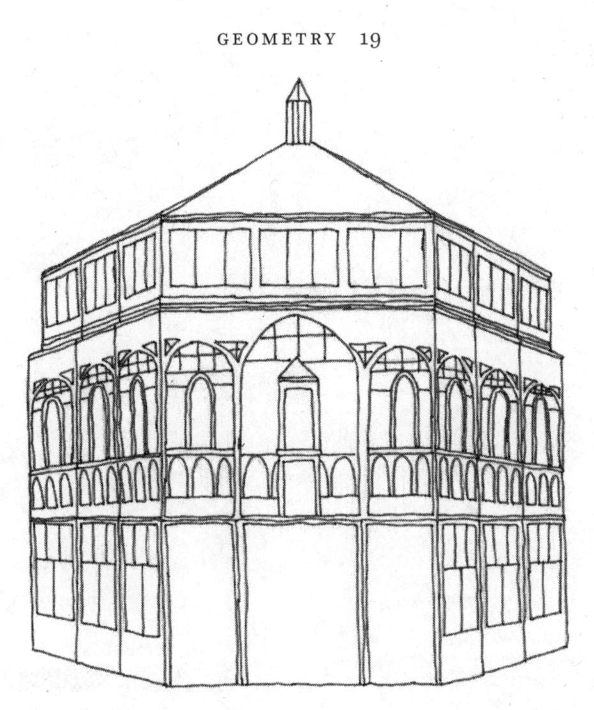

tion of self-similarity was an effective way to capture the notion of infinity.

Dalí found a sort of tiling to represent hidden infinities. Five hundred years earlier, Italian architect Fillipo Brunelleschi had found a geometric method to represent how objects appear. His 1415 painting of the baptistery in Florence, guided by a clever experiment with a mirror and a pinhole, was an early, perhaps the first, example of the Renaissance (re)discovery of perspective geometry.[5] Some art historians think ancient Greek and Roman artists had understood the geometry of perspective; others think their grasp of perspective was primitive. Medieval art often correlated the size of figures to their religious or political importance, ignoring their relative positions. Brunelleschi's notion was that paintings should present objects as they appear to us. Perspective geometry is the key.

The geometry of four dimensions, by contrast, does not appear to be grounded in our experiences, so is sometimes thought impossible to understand. A wonderful introduction is the mathematician Thomas Banchoff's book *Beyond the Third Dimension*.[6] Among many ways to understand the four-dimensional cube, or *hypercube*, Banchoff describes the method of unfolding. A cube (that is, the surface of a cube, not its insides) can be unfolded into six squares, shown in the left image. A hypercube, Banchoff shows, can be unfolded into eight cubes, shown on the right image. Why does the boundary of the hypercube consist of eight cubes? We'll explain this in the appendix, but maybe you'll be satisfied by this sequence: the boundary of a (two-dimensional) square is four (one-dimensional) line segments, and the boundary of a (three-dimensional) cube is six squares, so the boundary of a (four-dimensional) hypercube is eight cubes.

Dalí's fascination with science and mathematics is well-known; Banchoff met Dalí to discuss four-dimensional geometry and carried on a correspondence with him. Art and geometry are natural allies. Dalí's 1954 *Crucifixion (Corpus Hypercubus)*, sketched on the next page, uses an unfolded hypercube for a cross.[7]

How's this for a reason to study geometry? You might get to talk with Dalí. Well, not Dalí, who died in 1989, but maybe you'll get to talk with some other famous person. I got to spend time backstage at the Schubert Theater in New Haven with the comedian Demetri Martin, known for his contributions to *The Daily Show*, because he was a student in my fractal geometry class.

* * *

For our last example, we begin about twenty-three hundred years ago in Alexandria, home of the Greek mathematician Euclid. Because its elements were assembled by him, the geometry we

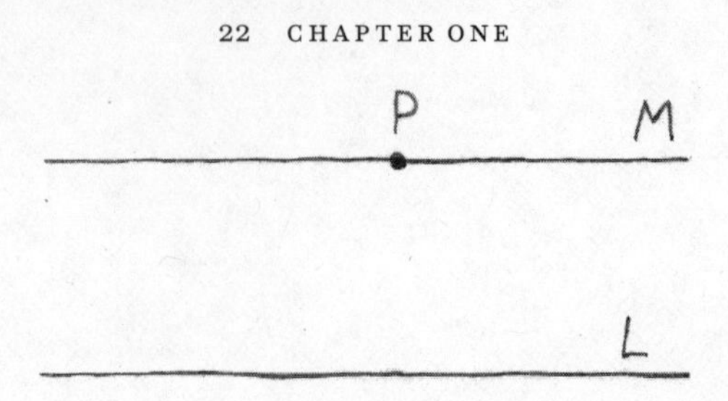

learned in high school is called Euclidean geometry. Every part of this geometry—the constructions, the thicket of theorems about triangles, all of it—follows from five assumptions, called *Euclid's postulates*. The first four are simple and easy to believe: any pair of points can be connected by a line segment, a line segment can be extended in the same direction as long as we like, every line segment is the radius of a circle, and all right angles are equal.

The fifth, called the *parallel postulate*, is different. It says for any point P that does not lie on a line L, there is exactly one line M passing through P that never touches the line L. We say M is parallel to L. This makes sense: if we tilt the line M even a tiny amount in either direction, eventually M will cross L.

The parallel postulate is different from, more complicated than, Euclid's other four postulates. Well into the nineteenth century, some mathematicians tried to prove that the parallel postulate was a consequence of the other four. These efforts were doomed to fail, because there are geometries, called *non-Euclidean*, in which the parallel postulate is false.[8]

In his 1959 woodcut *Circle Limit III*, sketched on the next page, M. C. Escher used non-Euclidean geometry.[9] For some time Escher had experimented with ways to represent the infinite in a finite region. Checkerboard tilings suggest patterns that continue to infinity, but Escher sought something better than suggestions.

The Poincaré disk, developed by the brilliant French mathe-

matician Henri Poincaré, provided the answer. Here the entire infinite plane is compressed to the inside of a disk: the ruler shrinks as you get closer to the edge of the disk (closer, at least, in the sense of familiar Euclidean geometry). Measured with this Poincaré ruler, the distance from the center of the disk to its edge is in fact infinite. And the area of the Poincaré disk is infinite. And those aren't the only differences from Euclid. In the Poincaré disk straight lines appear in two forms: as straight lines through the center of the disk but also as arcs of circles that intersect the boundary at right angles.

But wait, how are arcs of circles straight lines? This is an example of one of the central ways math grows. Take an idea in one setting—straight lines in the plane, for example—and figure out how to transport it to another setting. What aspect of straight lines can we generalize? In Euclidean geometry, a straight line is the shortest distance between two points. Let's use that. You probably already know one example of this generalization if you've flown long distances. A great circle on a sphere is any circle whose center is the center of the sphere. Longitude lines are arcs of great circles, while the only latitude lines that are arcs of great circles lie on the equator. On the sphere, the shortest distance between two points is an arc of a great circle that passes through those points. Stretch a rubber band between two points on a ball. The rubber band is the shortest path *on the sphere* between those points. And this shortest path will be an arc of a great circle.

To minimize travel time and fuel consumption, long-distance airline routes are arcs of great circles on the earth. For example, the latitude of Los Angeles is 34.1°N and the latitude of Moscow is 55.8°N, yet the airline route between these cities passes over northern Greenland, about 70°N.

Back to the Poincaré disk. With distances as measured by the Poincaré ruler, the shortest path between two points is either a segment of the disk diameter or the arc of a circle perpendicular to the disk boundary between the two points. In this sense, these are the straight lines in the Poincaré disk.

Why is this geometry non-Euclidean? In the Poincaré disk, we see that for a point P not on a given line L, there are many—in fact, infinitely many—lines parallel to L (in other words, lines that never intersect L) that pass through P. The lines M and M' pictured on the next page are two examples.

You may remember in geometry class learning a bunch of theorems (side-angle-side might ring a bell) that guarantee two triangles are congruent, that is, they have the same shape (are similar) and also the same size. In the Poincaré disk, it's a bit

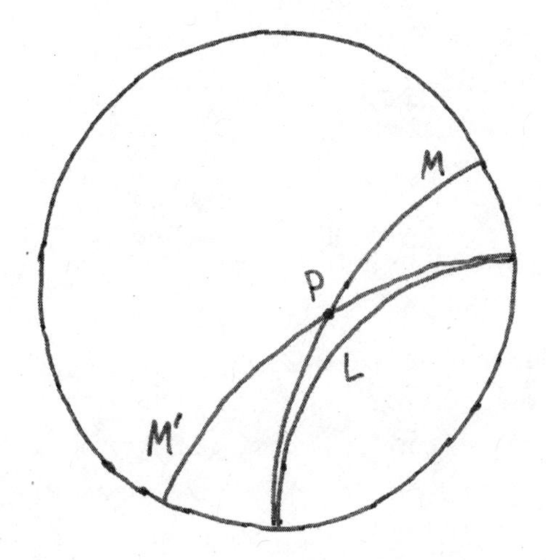

easier: similar triangles are always congruent. Consequently, in Escher's picture what we see as ever-smaller fishes approaching the edge of the disk are all the same size when measured with the Poincaré ruler.

Escher saw a picture of the Poincaré disk in a paper by the mathematician H. S. M. Coxeter, and the two corresponded about non-Euclidean geometry. Though Escher took some artistic liberties in *Circle Limit III*—as Coxeter pointed out, the curves in the image are not quite non-Euclidean lines—its mathematical inspiration is sound.

A comment about my sketch. In Escher's image, the little fishes continue all the way to the boundary. Not exactly, because that would require infinitely many fish, but Escher continues much farther than I have. And I must mention the unforgiving process he employed. I noted earlier the meticulous labor behind the Sierpinski gasket cathedral tiling. Yet if one of those tiles was carved incorrectly, the artisan could just carve another piece, assuming an adequate supply of stone. Escher's art is a woodcut; he carved all of these fishes into a single block. A mistake could ruin the whole carving, not just a tiny section. Think of this as you seek inspiration for patience.

* * *

Geometry is a way to organize our models of the world, its shapes and dynamics. But isn't this all contingent, balanced on a knife's edge? Could our models have turned out very differently? If the fractal geometry of Mandelbrot had been discovered before the geometry of Euclid, would manufacture be the same? If you think the question is far-fetched, consider the iterated branching of our pulmonary, circulatory, and nervous systems, or the recursive folding of our DNA, or the large surface area and small volume of our lungs and our digestive tract. Evolution has discovered and uses fractal geometry. If people had looked more closely at the geometry of nature, rather than emulating the "celestial perfection" imposed by the church's interpretation of the works of Euclid and Aristotle, our constructions could be very different now.

Do the significantly different cosmologies of various cultures reflect different perceptions and geometries, or alternate paths in their historical narratives? Unless there were only one geometry, only one story—only one world—we should not expect the same categories to grid our views of the universe.

This, really, is the beginning of our main point. Could the world be different than we think? Is it different? Must it be only one thing, or can it be many? If we view the world in one way, does this forever bar us from all others? In the many worlds model of quantum mechanics, described clearly in Sean Carroll's beautiful book *Something Deeply Hidden*, every observation of every particle splits the universe into branches, one in which each measurement outcome occurs, and communication between these branches is impossible.[10] So in physics we have a model in which one choice bars us from all others. Does this separation percolate up into the world of people and clouds and cats? We'll think some about this as we go along.

This brings us again to grief, a response to irreversible loss. Does the careful exploration of one geometry irreversibly color our understanding of the shapes of the world? In mathematical sciences the distance between dream and exploration usually is much smaller than it is in the experimental sciences. As in any science, you need to learn the background techniques. But in math you don't design experiments, assemble equipment, submit to an ethics review if you plan to use live subjects, run safety checks, then run the experiment, gather the data, and interpret the results. In math you just start to think. Okay, nowadays sometimes you write code and run simulations, but this too is mental, not physical, apart from typing the code into your computer. The worlds we explore are in our minds. This potential loss of other worlds, when focus on one blocks us from others, is a source of grief in the study of mathematics. Not grief of a magnitude comparable to that of losing a person or an animal, but nonetheless, I believe, the same flavor of emotion.

Now you may feel that this is silly. What, really, is lost? Can't we think in another direction whenever we want? In some sense we can, but once we've seen a new way to look at the world, we can't unsee it. I'll illustrate this point with an example from fractal geometry. If you aren't a fan of geometry, you can substitute some other complex, nuanced pursuit that you love.

For the moment, ignore the grid in the figure on the next page. Does the shape look complicated or simple? If you think it looks simple, you should be able to give an exact description of how to draw it. Can you?

Now look at the grid. Note that five of the squares are empty. It turns out this is almost all we need to know: keep track of these empty squares, and we can generate the shape. The process is

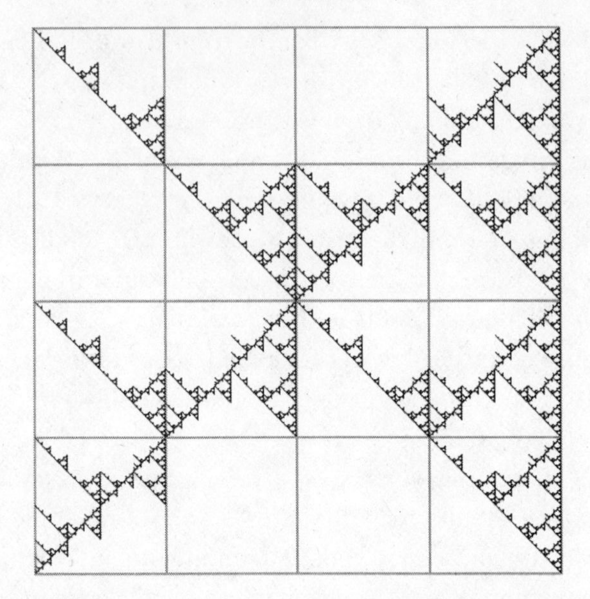

fairly simple. Start with a four-by-four grid. First, leave the five empty squares empty and completely fill the other eleven. This is the left image below. Second, shrink this image by one-half and place four copies, one in the lower left, one in the lower right, one in the upper left, and one in the upper right. This is the middle image. Third, cut out, from the middle image, the five large squares that are empty in the left image. This gives the right image.

Then repeat the second and third steps, starting each time

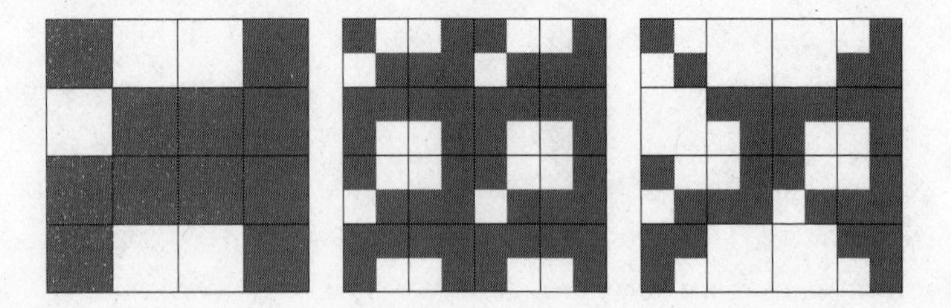

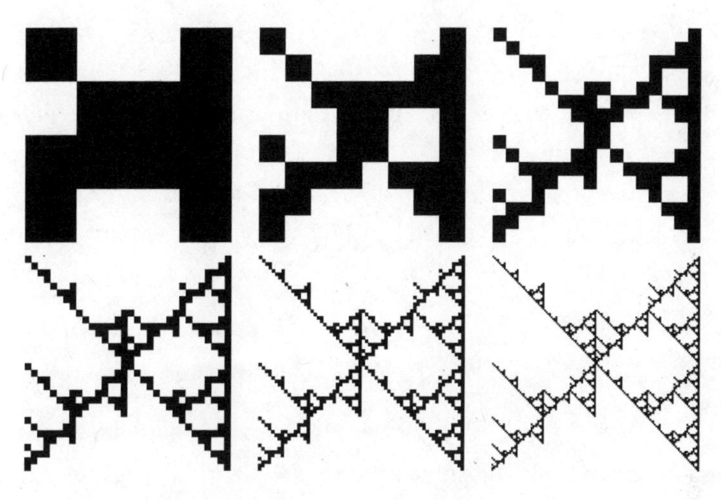

with the image just generated: take the final image from the previous iteration; shrink it by one-half; place copies in the lower left, the lower right, the upper left, and the upper right; and, again, empty the five large squares that were empty in the initial image. In the above figure, we see the initial four-by-four image and the results of the first five repetitions of this process. With each repetition, the shape gets closer to the image I first showed you. You may notice that small pieces resemble the overall shape. If you think this is a fractal, you're right.[11]

You might think of this as "fractal sculpture." Michelangelo is reputed to have said that every block of stone has a statue inside it. The task of the sculptor is to discover the statue. We've seen here that all we need to generate this fractal are the list of empty squares and a repeating procedure. The product may look pretty complicated, but from this point of view, it's quite simple. It should be no surprise that how complex something appears can depend on the tools we bring to analyze it.

Once you learn to recognize the fractal aspects of objects, your perception is changed forever. Over the years I've gotten dozens of emails from roommates of my students, all variations of a single complaint: "Every time we walk to class, my roommate spots a fern or a cloud or a crack in a sidewalk, and our conversation is interrupted by 'Here's a fractal,' 'There's a fractal.' Enough with the fractals already! You've ruined some good conversations." I stand accused of polluting history-major minds with geometry.

* * *

I do believe that once they are recognized, these patterns cannot go unnoticed. They change forever how the image of the world unfolds in our minds, change forever the categories of the models we build.

The first real example I experienced was in my high school geometry class. We'd spent some time in the study of compass and straight-edge constructions, puzzles much loved by the ancient Greeks. We'd learned how to divide a given line segment into two, or three, or four, or any number of equal size pieces. Then our teacher, Mr. (Ralph) Griffith, told us that the ancient Greeks had found three problems they could not solve: trisecting an angle (constructing an angle one-third the size of a given angle), squaring a circle (constructing a square with the same area as that of a given circle), and duplicating a cube (constructing a cube with volume twice that of a given cube).

Puzzling over these, I had an idea. Take the angle $\angle AOB$, the angle that starts at point A, follows the straight line to point O, and then the straight line to point B. (See the sketch on the next page.) To trisect the angle $\angle AOB$, I thought, just trisect the segment AB, the piece of the straight line that connects point A to point B. That is, find points C and D on AB so the length of BC equals the length of CD equals the length of DA. This we'd just learned to do. Then, I guessed, the angles $\angle AOD$, $\angle DOC$, and $\angle COB$ would be equal, and consequently, $\angle AOD$ would be

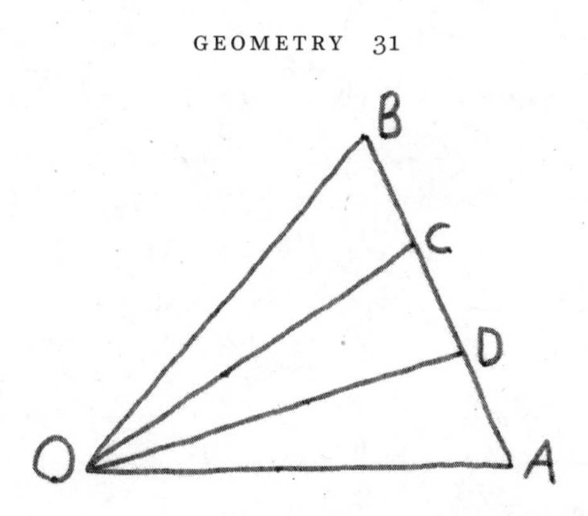

one-third of $\angle AOB$. That this simple idea—the first, really, that would pop into anyone's head—had somehow gone unnoticed for twenty centuries didn't strike me as odd or unlikely. That doubt didn't enter my thinking at all. I had a momentary glimpse of a newspaper headline: "Local Student Solves Two-Thousand-Year-Old Math Problem."

I showed my construction to Mr. Griffith. I had done a tiny drawing with a cheap compass. The angles looked close to equal with my protractor. Mr. Griffith did a larger drawing with a better compass. With his protractor the angles were seen to be different. Mr. Griffith hadn't said, "If it was that easy, don't you think someone would have figured this out two thousand years ago?" Rather, he was happy that I'd tried.

I'd assumed geometers just hadn't found the right approach. No, Mr. Griffith said, there are some problems that we can prove can't be solved. What? How can a problem be unsolvable? But even more astonishing, how can we know a problem can't be solved? Ever. The dizzying fact that there is a proof that some theorems cannot be proved I wouldn't learn for three years.[12] These impossibility of the three geometric constructions I wouldn't understand until many years later.[13] The math involved is complex and sophisticated—little wonder the ancient Greeks didn't figure it out.

I didn't know this when I was in high school. I'd known impossibilities in the physical world for as long as I'd known anything. I can't flap my arms and fly to the moon. Less silly targets also elude me: I am clumsy, without grace or even competence. But geometry—that some things are impossible in geometry was unsettling. How could there be problems in geometry that couldn't be solved if one worked hard enough? Something seemed seriously wrong with a universe in which this is true.

I asked Mr. Griffith how one could prove that a mathematical construction is impossible. He didn't try to explain the angle trisection proof. Instead, he showed me a proof that the square root of 2 cannot be written as a ratio of whole numbers, another result that shook the foundations of Greek geometry. The proof is clean, clear, elegant. (And involves a little bit of algebra; you can find it in the appendix.) Mr. Griffith's gentle exposition, guiding me to fill in some steps, made me so happy.

That night, thinking over the beautiful proof, I realized that geometry has boundaries too. This bothered me for about ten minutes. Then I realized that these boundaries made geometry even more interesting. Just how much more interesting I wouldn't realize for years and still don't understand fully. Turns out what I'd thought was a map of the whole world was just a tiny corner of the map.

The next day on the walk to school I ran through the steps of the proof again. The pieces still fit together beautifully, but I'd lost the initial thrill of recognition, what some people call the aha moment. The moment when observations or ideas rearrange themselves and you discern a new pattern, crystal clear now but invisible before. The new arrangement of ideas will stay with you, but the aha won't. For a given pattern, you get at most one aha.

When I taught fractal geometry, the second lecture contained the biggest aha moment of the semester. The essence of that moment is a sequence of pictures that turn a sketch of a cat into a Sierpinski gasket.[14] Weeks later, when we explored other topics, some fairly complicated, students complained that they wanted more surprises like that of the second day.

On the other hand, ignoring the new arrangement (fractals) in favor of the old can be a challenge. I expect I've heard a hundred variants of "I miss the days when I'd look at a tree and think only that it is pretty. Now I can't help but look for transformations that make the tree shape." It's a good thing, these students suggest, that John Muir (or Rachel Carson or Edward Abbey) didn't know about fractals. Short of substantial head trauma, there's no way to unsee a new idea.

Martin Gardner, whose "Mathematical Games" column appeared in *Scientific American* most months from January 1957 to June 1986, wrote a book of math problems called *aha! Insight*.[15] The problems are clever, with "outside the box" solutions, entertaining if you like math puzzles. Buth these small-scale aha moments don't effect irreversible changes in how we view the world.

Or do they? Might they reshape, not how we view the whole world, but how we let our imaginations steer us clear of straightforward, but overcomplicated, approaches. I'll illustrate this with a puzzle called the *bumblebee problem*.

Imagine a straight railroad track that runs east to west and is 50 miles long. At the west end of the track is a train engine that will travel east at 30 miles per hour; at the east end of the track is a train engine that will travel west at 20 mph. Both begin to travel at noon. At noon, a bumblebee starts from the front of the eastbound engine and flies east at 70 mph. When it reaches the westbound engine, it turns around and flies west until it reaches the eastbound engine, at which point it turns again, flying east until it reaches the westbound train, and so on. (See the drawing on the next page.) Both engines are moving all the while. The question is: how far will the bumblebee fly before it is caught between the colliding engines?

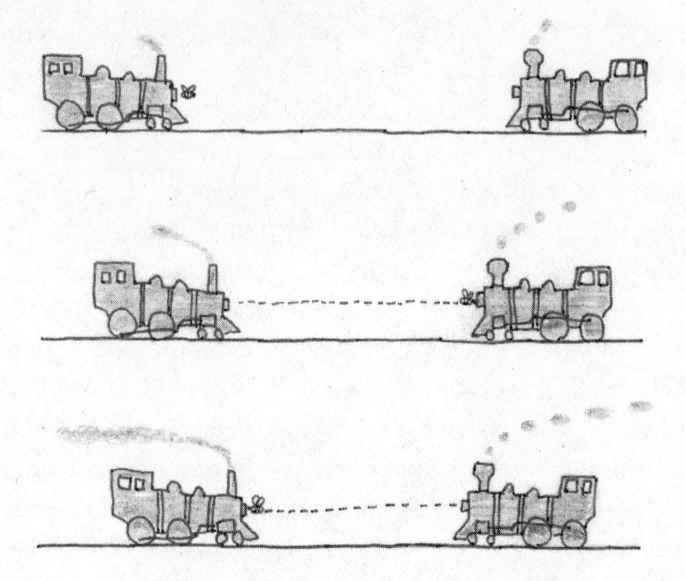

I was introduced to the bumblebee problem when I was in the seventh grade. That year two NASA engineers visited St. Albans junior high school. Recruiting bright science and math students, I imagine. Over lunch period my science teacher found me and asked me to talk with the engineers before their presentation that afternoon. One of them explained the bumblebee problem and asked if I could solve it.

Okay, when the bumblebee flies toward the westbound train the relative speed of the bumblebee and that train is 70 + 20 = 90 mph. The distance both must travel is 50 miles, so the time until the bumblebee encounters the westbound train is 50/90 hours, or 33⅓ minutes. I know the bumblebee's speed and the travel time, so I can find the distance it traveled: distance = speed × time.

I can also figure out how far each engine traveled in that time. Subtract the sum of those distances from the original 50 miles, and that's how far the bee and the eastbound train must travel until they meet. I'd learned about geometric series, infinite addition problems where each term is a fixed multiple of the one before it. For example, 1 + ½ + ¼ + ⅛ + . . . is a geometric series where the fixed multiple is ½. If you can recognize the pattern and find the fixed multiple, there is a simple formula that will give the sum. But to find the fixed multiple for the bumblebee problem was not so easy. With paper and pencil I could work this out in an hour probably. So I knew a way to do the problem, and it was a mess.

But I didn't have paper and pencil, and the engineer had just asked if I could solve it. He hadn't said, "Think about it for a bit." He seemed to want an answer now. What was I missing? What if I ignore the bee and think just of the trains? They've got 50 miles to travel, and their relative speed is 50 mph, so the trains will meet in an hour. But wait. The bee flies 70 mph, so in an hour it will fly 70 miles. Is that all there is to the problem? No complicated geometric series? So I said, "70 miles." Both engineers smiled. One told me to look them up when I finished college. I didn't look them up. Where would I be if I had?

Some years earlier the bumblebee problem had been posed to the brilliant mathematician John von Neumann. Von Neumann worked on the Manhattan Project at Los Alamos, was a colleague of Einstein at the Institute for Advanced Study in Princeton, and was a central figure in the design of modern computers. Benoit Mandelbrot, the inventor of fractal geometry, was von Neumann's last postdoc at Princeton. From an early age, von Neumann could multiply two eight-digit numbers in his head. When told the bumblebee problem, von Neumann stared into space for a couple of seconds and gave the answer. "So you saw the trick," the questioner said. "What trick?" von Neumann replied. "I summed the series." In this case, great technical facility kept von Neumann from seeing the simpler approach.

What changed when I thought about this on the walk home

from school is that now I knew some problems can be solved in several ways, and the first approach that comes to mind might be unnecessarily complicated.

The little aha moment, what a scientist might call the local aha moment, was finding the trick to solve this problem. The big, or global, aha moment was the realization that the first approach to a problem is not necessarily the one you want to pursue. Until then, as soon as I saw a strategy to solve a problem, I jumped in and started to work. Even now, forty-five years later, finding a first strategy gives license to take a breath and let the imagination dance around the problem. Is there another approach? In class when we began a complicated problem, after we'd found a first approach, I'd ask my students to find another. "Why?" some wondered. Because we might find a simpler way to solve the problem, and because occasionally a comparison of two approaches can reveal a previously unseen aspect of the problem. Once I'd turned this corner, I couldn't go back. I think, I hope, a few of my students got the point, though the majority seemed unconvinced that they should spend time looking for a second approach. Many resisted this new perspective on problem-solving.

So does nostalgia for familiar modes of thought mean that we should not learn new things? Of course we should learn new things. Old ways of seeing close because new ways of seeing open. This is how we must move through the world. But the inevitability of closing doors should not provoke us to we embrace ignorance of alternatives and explanation. Even though our lives of necessity involve repeated loss, unexamined loss is unbearably worse.

Now you may feel little surprise that geometry can provide insight into how we understand nature. But the goal of this book is different, or at least differently focused. I aim to show that geometry can provide insight into how we understand our sense of loss. To foreshadow this approach, we'll now show that geometry can lead to surprises in how we can interpret literature.

The story "In Circular Ruins," written by Jorge Luis Borges in 1940, occupies just over five pages in his collection *Labyrinths*.[16] It is a remarkable fiction of a person who dreams another person into the waking world, maybe. If you haven't done so already, you should read the story.

We know that mathematics was familiar to Borges.[17] Paradoxes and puzzles, especially those that involve infinity, fascinated him.[18] We can find manifestations of geometry in visual art in a more straightforward way than we can those in literature. This is no surprise: the medium of visual art is shapes we see and shapes we don't see, positive and negative space. We can see all of a painting at once, or we can choose to direct our gaze to a particular area of the canvas. Literature, on the other hand, is, like music, grasped sequentially. Unless we understand a piece so thoroughly that we can hold it in our minds in its entirety, we get little bits, one piece after another. Seeing larger-scale patterns requires memory and deduction. Let's focus on literature. Because we have only a limited amount of information—all the words of a story, maybe the body of the author's work, maybe some knowledge of the circumstances of the author's life—we must make inferences. We can't ask Borges if our interpretation is correct. All we can do is analyze our conjectures and our supporting reasons.

"In Circular Ruins" is the story of a man who travels from his home among the infinite villages on a violent mountainside downriver to a ruined circular temple. His goal is to dream a person complete in every detail and to insert him into the world. The first attempt, based on dreaming a crowd of students and selecting a promising member, failed. But the second, a year spent dreaming anatomically and building a person organ by organ, succeeded. The god of the temple, who is at once a horse and a tiger and a bull and a rose and a tempest (a good example of how Borges deployed his imagination—a simple list twists in an unexpected way and we're left breathless) and whose name is Fire, has brought the dreamed person into the world. Only Fire and the dreamer know the dreamed one is a phantom. The dreamer trained the dreamed one for two years, then removed the dreamed one's memory of his training and sent him upriver to a second ruined temple, also a temple to Fire. Some time later, the dreamer heard of a magician in an upriver temple who was untouched by Fire. Would the dreamed one realize he was dreamed? Then Fire surrounded the dreamer's temple, but had no effect on him. The dreamer realized that he was himself dreamed.

* * *

This arrangement, where the dreamer of the dreamed one is himself dreamed, is what we want to unpack. Can geometry help us find something new in this story, something hidden? That the dreamer is dreamed can lead to several geometries.

1. *One-off.* The dreamer and the dreamed one are the whole story. There's nothing more.
2. *Dreamers all the way down.* The dreamer and the dreamed one are part of an infinite sequence of dreamers dreaming other dreamers.
3. *Round and round.* The dreamer and the dreamed one are the same, and time is circular, allowing for some noise, some variation, on successive trips around the circle.
4. *Round with a (Möbius) twist.* The dreamer dreams the dreamed one, and the dreamed one dreams the dreamer.

We'll consider each in turn. Keep in mind that we are making inferences about how we think each geometry fits Borges's imagination. Other readers may reach other conclusions.

One-off. There is an ur-dreamer who dreams the main character of the story. So the ur-dreamer dreams the dreamer, who in turn dreams the dreamed one. But the dreamed one dreams . . . no one? That's a brutal, inelegant, asymmetrical structure for a story from the imagination of Borges. We cannot believe this most graceful storyteller would set his beautiful prose in the service of such a mundane idea.

Dreamers all the way down. There is an old story of a scientist (Bertrand Russell, Carl Sagan—many scientists have been inserted into this story) who, after a lecture on astronomy, is told by a member of the audience that in fact the world rests on the backs of four immense elephants who stand on the back of an immense turtle. Smiling gently, the scientist asks, "And on what does the turtle stand?" "You are very clever, but it's turtles all the way down," is the reply.

We don't know if Borges was familiar with this story, but in various forms it's been around since at least the mid-nineteenth century. And Borges certainly was familiar with the basic arithmetic of infinities.[19] We find two reasons this model is unlikely to underlie Borges's story.

First is what we'll call Borges's sense of fair play. If there are infinitely many dreamers and dreamed ones, how could he tell us just of two and give no hint of the others? No sketch of their stories, of how one is related to another, nothing. What would be the point of talking about only two characters if infinitely many others are forever hidden, unknowable even in principle? Certainly, Borges is familiar with Occam's razor, commonly stated as "The simplest explanation is the most likely," but William of Occam's original statement (translated from Latin into English) was, "Entities must not be multiplied beyond necessity." This is pretty much on the nose for the application we're making: an infinite collection of extraneous characters is, well, infinitely far from a construction consistent with Occam's razor.[20]

Second, the timing is problematic. The dreamer of "In Circular Ruins" persists in his illusion of reality longer than does the person he dreams. Going forward, these times become ever shorter, eventually becoming impossibly short. Going backward, these times become ever longer, and this, too, is a problem.

Round and round. Could it be that the dreamer and the dreamed one are the same? This would turn the story into a circle.[21] The notion of circular time was familiar to Borges. Indeed, one of his essays is titled "Circular Time."[22] There he raises an essential consideration: "the concept of similar but not identical cycles." Each time around the circle can reveal modest variations from the previous cycle. But how modest? That the dreamer and the dreamed one have different experiences before entering their ruined temples is a modest difference. About the dreamer we read, "If someone had asked him his own name or any trait of his previous life, he would not have been able to answer." About the dreamed one, "He [the dreamer] instilled into him [the dreamed one] complete oblivion of his years of apprenticeship." Neither knows much of what happened before he entered his temple, so this difference is inconsequential to how the stories unfold.

On the other hand, the dreamer and the dreamed one spend substantially different lengths of time thinking they are real. This difference could have a considerable impact on their stories—how could it not?—so we think this distinction is important. The dreamer and the dreamed one cannot be the same. Borges's story does not support a circular geometry.

Round with a twist. So far we have assumed that there are exactly two distinct characters, the dreamer and the dreamed

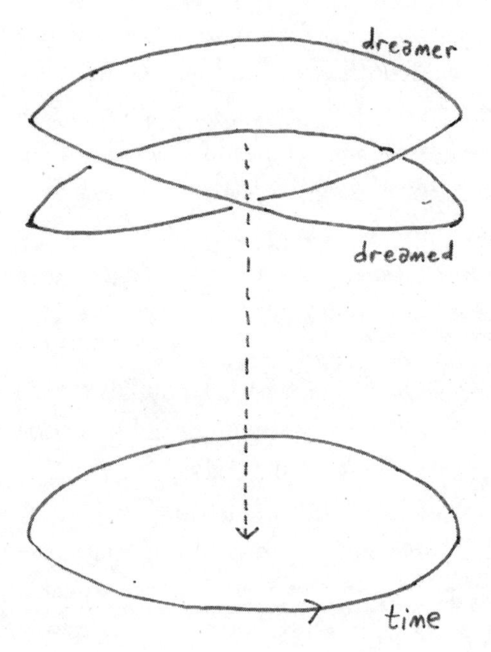

one. In our first scenario, we raised the issue of symmetry, specifically, why doesn't the dreamed one dream a person? To avoid an infinite collection of dreamers, with the problems mentioned in the second scenario, the apparent solution is a loop: the dreamer dreams the dreamed one, and the dreamed one dreams the dreamer. My sketch represents this schematically using the geometry of a shape called a Möbius band, which has only one edge and only one side. (You may recall Möbius bands from early math, or maybe art, classes: take a long, thin strip of paper, give one end a half twist, and glue the short ends together.) We can project every point of the Möbius band to its time, represented as a circle, so that each character can dream the other. As we mentioned in the third scenario, Borges was familiar with the notion of circular time. Now, we don't need the whole Möbius band: above any given point on the circle (that is, at a particular time), we need only two points, one representing the dreamer at that time, the other representing the dreamed one at that time. Technically, the story constitutes the boundary of the Möbius band. To be sure, during the time in which the dreamer tries his first experiment in dreaming a person into reality, the dreamed one is a potential dream rather than an actual dream. We think this is a simple variation, within our "close enough" criterion for finding geometry in the structure of literature.

Geometric considerations have got us to this place. Although foreshadowed in the story, the final sentence, "With relief, with humiliation, with terror, he understood that he too was a mere appearance, dreamt by another," shocking as it is the first time we read it, is not the end of the story. The story has no end: it is an autonomous universe, self-generating, locked in the same cycle, though minor variations can occur. Borges speculated on, and presented some calculations about, the cyclic nature of finite universes in his essay "The Doctrine of Cycles." Geometry has lead us to deduce that "In Circular Ruins" is a narrative expression of this eternal recurrence.

We'll end this chapter with a lesson that eternal recurrence provides us about the humility we should have when we toss around the word "infinity." Rather than recurrence in time, we'll talk about copies of ourselves, with variations, in space, if space is of infinite extent. We see only a small part of infinite space: the *observable universe* is what can be seen from earth now. Time since the Big Bang is about fourteen billion years. Add in the effects of early cosmic inflation and the observable universe has a diameter of about ninety-three billion light-years. (Why not twenty-eight billion light-years? That's where inflation comes in.)

If space is of infinite extent, then it's filled with many—in fact, infinitely many—parallel universes, observable perhaps for other worlds a long way from us, but not observable by us.

This whole collection is called the *multiverse*. The only other assumption is that at very large scales (far, far larger than the scale of clusters of galaxies, for example) matter is approximately uniformly distributed in space. Measurements of the observable universe hint at this. As Max Tegmark explains in his *Scientific American* article "Parallel Universes," and in more detail in his book *Our Mathematical Universe*, from these two assumptions (infinite space, uniform distribution of matter), within a distance of about $10^{10^{118}}$ meters (a really big number[23]) from us, we will find another parallel universe *identical* to ours.[24] Identical in every way—every quantum state the same as in our observable universe. There we would find another copy of you: same brain structure, neural connections, memories, all identical to yours. Moreover, there are infinitely many copies of this identical universe, infinitely many copies of you. But that isn't the point I want to make here.

First, a comment about what some cosmologists (including Tegmark) think are in these parallel universes. Energy fluctuations soon after the Big Bang would result in a degree of randomness in the initial conditions, so if we look far enough we'll see every possible arrangement not forbidden by the laws of physics.

If space is truly infinite, then at immense distances there are parallel universes identical to our observable universe, except that in this version . . . in class I'd hold a piece of chalk above the table and ask, "Will I drop the chalk, or not?" Some years I would, some years I wouldn't. Whichever I did, I'd follow it with this: "Unimaginably far away there is another identical copy of our observable universe, except in that version I'll do the opposite with the piece of chalk."

But it's not just me with this stupid piece of chalk. Think of every possible variation in your life. For each minute change there is a parallel universe where everything is the same, except for that change and its consequences. And not just your life, but the lives of everyone. And not just a single change, but all combinations of changes. And not changes just in our lives, but in every blade of grass, every grain of sand, everything on every planet, every swirl of plasma in every star, every pinwheel of a galaxy spread across the deep dark void.

Then think of this. Even though we know how to manipulate the mathematics of infinities, how could we think that we *understand* the infinite? I'd have more success teaching the field equations of general relativity to one of my cats. "Okay, Bippety, here's how to define the covariant derivative." Not going to work.

Three last remarks before we move on. Tegmark's book *Our Mathematical Universe* puts forward the view that the universe isn't just well modeled by mathematics. Rather, the universe *is* mathematics. This view has not been universally embraced by the scientific community, and while I can't yet say that I believe it, the notion is beguiling.

In this dance we can trade partners, swap infinite space for infinite time. One complication is implied by the second law of thermodynamics, which states that in a closed system, entropy, a measure of the disorder of a system, tends to increase or stay the same. Here's the complication: the increase of entropy implies that it was lower in the past, and very much lower at the Big Bang. Low-entropy configurations are less common than high-entropy configurations, because there are many more ways for a system to be disordered than ways for that system to be ordered. For example, there's only one way for your coffee cup to be unbroken (high order, low disorder, low entropy), but there are an awful lot of ways for your coffee cup to be broken (high disorder, high entropy). So why, or how, was entropy so low at the Big Bang? Ludwig Boltzmann, architect of the mathematical definition of entropy and of this statistical-mechanics formulation of thermodynamics based on counting the number of states of a given entropy, took a bold approach to the problem. Not to this problem exactly, because Boltzmann died in 1906, many years before George Gamow and Ralph Alpher published calculations for their Big Bang model.[25] Here's Boltzmann's approach. Almost always the universe is in a high-entropy state, a cold diffuse approximately uniform distribution of thermal radiation. But if time is infinite, then eventually random fluctuations will generate lower-entropy pockets. And no matter how unlikely this may seem, one of these fluctuations will be a very low-entropy clump that to us will look like the Big Bang in our past. And (probably you anticipate this) not just one fluctuation, but infinitely many, and with all possible variations.

Boltzmann's approach is described in Sean Carroll's book *From Eternity to Here*.[26] I am smitten with this argument, though Carroll later points out some objections to Boltzmann's approach. (Look for "Boltzmann brains" in Carroll's book to see a cute formulation of one of these objections.)

Carroll also presents another explanation that is fascinating though not yet embraced by all cosmologists. In the realm of quantum gravity (still *very* far from being understood completely), space-time itself can fluctuate. Some of these fluctuations can become *baby universes* that break off from another universe, begin in low-entropy states, and inflate to form another universe. The randomness of the fluctuations that give rise to these baby universes can produce an immense variety of initial conditions, so . . . by now you can fill in the blank.

You may feel that we've wandered a long way from thoughts of grief, but we haven't. In moments of wild panic at the irreversible loss of a loved one, I have found a bit of comfort in the thought that some parallel universe or some distant future contains a version of me who has found a way to dampen the fire of grief.[27] I wish I could talk with that guy, but of course I can't. But maybe, eventually, I'll be that person, or maybe you will be. Then we can have a very useful conversation.

2
Grief

How we gentle our losses into paler ghosts.

PETER HELLER

Some people think grief is nothing other than sadness on steroids. But I believe grief differs from sadness both psychologically and philosophically. I am sad when the day I'd planned a walk to the park for a pleasant hour with the new Murakami novel begins with heavy rain that lasts throughout the day. This is an inconvenience, a disappointment, but it does not signal an impossibility: there will be other days, other walks, and (I hope) other new novels by Murakami.

On the other hand, when I stood beside my mother's coffin I knew—*knew*—she was gone. No more conversations, never again to be enfolded in a mother's hug, a moment of comfort and security from earliest childhood until the last goodbye, a few weeks before a stroke, unexpectedly, ended her life. This is grief, and it is irreversible.

Shortly after Mom died, one of my students had asked me to write a brief piece about gravity to be a part of a class project that she was assembling. "But really, write anything you want, so long as you mention gravity." I wrote about gravity and Mom:

> Gravity holds my feet on the ground. Gravity keeps the earth traveling around the sun, the sun dancing around the galaxy, the galaxy threading through the Local Group, and on and on.

Gravity pulls rain out of the sky. And snowflakes. And leaves in autumn. And tears from my eyes when I knew you really are gone. Where did you go? Why can't I see you anymore? Why can't I remember your face?

Who can ignore gravity? Not birds. They are just better at pushing against it than we are. Fish can ignore gravity. Imagine life with a swim bladder. Push inside in some subtle way, then a ballet on clouds, through clouds. Tap down on a mountain top, then push off into the sky.

This is how to learn that the distance between here and there is the answer to the wrong question. Can I tell you the right question? Let's see.

I thought gravity pulled my mind into the past, stuck in memories. But now I know I can't trust memories. Some are invented, all are edited. The whole web of who I am—what I've seen and done, what skills I've found—is nothing but fog.

Gravity pulls me to the future, bits of me falling off along the way. Each of us disappears into the mist of the possible. In our minds, time is gravity's other side.

In the last moments before I vanish, will I see you again? Only one moment, one glimpse, is all I want. All I need. As my memory dissolves, is there no chance to see you, touch your face, hold your hand, see my face reflected in your eyes? Why can't I stop weeping? Why can't I find my breath? Everything is so small and dark. I hope I see you before I finish writing thi

I had no warning that Mom was sick. My wife Jean and I had visited Mom and Dad a few weeks earlier. The morning of the day she died, Mom and I had traded emails. That evening, while she and Dad watched the local news, Mom tried to get up from the couch. She couldn't stand. Tried again, still couldn't get up. Dad took her left hand but she didn't squeeze back. The left side of her face had begun to sag. Dad said, "I'll call 911." Mom said, "No, call Steve." "What's Steve's number." "One, two, three, four," and she was gone. Late that night Steve called with the news. Terribly shaken, we headed to West Virginia for the funeral.

We were unprepared. Little about Mom had changed over the last decade of her life. Her hair turned white, she moved more slowly, cut back on the hundreds and hundreds of Christmas cookies she baked, most to give to neighbors and to family. But she still was Mom, sweet and smart and happiest when all the family was in the house. Then she was gone. Then Dad was alone. His wife of over sixty years never again would sit across the table from him at mealtime, never again sit on the couch and read the newspaper while he sat in his rocking chair and watched a cowboy movie. I'd never again see Mom's name in my email inbox, never again hear her voice on the phone, never again sit up late into the night at the kitchen table with her and be amazed time and again when she saw perspectives that I did not. All this is gone. I was unprepared and the heartbreak was staggering. She's been gone a decade now and this still hurts. This is qualitatively different from sadness.

To emphasize the difference between this grief and the sadness of the rainy day, let's think for a moment about what losses will result from the rained-out walk. Now certainly if I miss a walk to the park, I will miss particulars of that walk. On a later walk I'll see different flowers in bloom, a different fullness of leaves on trees, different dogs at play, different birds in the sky. I'll read other pages, maybe another book. The intervening days will add experiences I wouldn't have had yet had I taken the earlier walk. These experiences would change my model of the world, and the context in which I interpret what I read, what I see at the park. But usually such changes are minor, not accompanied by a sense of loss. This is not a door closed; it is a slight shift in perception. And so far as I can tell, a shift without noticeable effect.

If you embraced the lessons of what's called chaos theory,

you may wonder how we can say that a slight shift might not have a noticeable effect. The basic ideas of chaos theory were developed first by Henri Poincaré in the late nineteenth century, but forgotten and rediscovered several times in the twentieth century, untill finally Robert May's 1976 paper on population biology and James Gleick's 1987 book *Chaos: Making a New Science* brought the topic into popular culture.[1] How popular? It was a central theme of the "Time and Punishment" segment of the "Treehouse of Horror V" episode of *The Simpsons*. And *The Simpsons* is an important, maybe the most important, arbiter of contemporary culture.

One of the characteristics of chaos is that small changes can have large effects or, as Abraham Simpson said, "If you ever go back in time, don't step on anything because *even the tiniest change can alter the future in ways you can't imagine*." In mathematics and science, this is called *sensitive dependence on initial conditions*. In class I'd start with a more technical definition, then give this restatement. Then show the "Time and Punishment" segment. As near as I could tell, most of my students thought that *The Simpsons* writers had used my characterization of sensitive dependence on initial conditions, when in fact I'd used theirs.

Sensitive dependence on initial conditions is also called the *butterfly effect*, sometimes summarized as "a butterfly flaps its wings in Boston and we get a tornado in Texas." (Adjust locations to taste.) However, this isn't what sensitive dependence on initial conditions means. My colleague Dave Peak points out that the tiny energy of a butterfly wing flap cannot organize itself into the immense energy of a tornado. In fact, this was mentioned in Poincaré's original formulation:

> We see that great disturbances are generally produced in regions where the atmosphere is in unstable equilibrium. The meteo-

> rologists see very well that the equilibrium is unstable, that a cyclone will be formed somewhere, but exactly where they are not in a position to say; a tenth of a degree more or less at any given point, and the cyclone will burst here and not there, and extend its ravages over districts it would otherwise have spared. If they had been aware of this tenth of a degree, they could have known it beforehand, but the observations were neither sufficiently comprehensive nor sufficiently precise, and that is the reason why it all seems due to the intervention of chance.

Small changes may not *cause* large differences, but small changes, invisible because of our inability to measure exactly, can *mask our ability to predict* whether, when, and where large differences can occur. Chaos is about the breakdown of our ability to forecast for more than a short time.

This is why chaos isn't so relevant to the distinction between sadness and grief. Grief is not about prediction; anticipation (usually) is not irreversible, and so (usually) it is not grief. In 2002 my brother Steve was diagnosed with CLL, chronic lymphocytic leukemia. Since then, his disease has been followed at the James Cancer Hospital in Columbus, Ohio. In January 2010 Steve collapsed—very fortunately, while he was at the James for a scan. His white cell count had risen to sixty times the normal maximum, his hemoglobin had dropped below a quarter of the normal minimum, and his kidneys had failed. He was rushed to the ER, then the ICU, where doctors started dialysis and put him on a ventilator. Until the Red Cross found a match for transfusions, his blood was removed, centrifuged to pull off most of the white cells, rehydrated with plasma, and returned to him. Steve began to thrash around, his movements using oxygen he couldn't spare, so his doctors administered a paralytic. They told his wife Kim and my sister Linda, who was at the James by then, to prepare themselves and to make the necessary phone calls. We did not expect he would survive the night.

All that night my skin buzzed. Memories of years with Steve and Linda crowded together, so much that even now they may remain cut, mixed, and pasted together into strings unrelated to the sequences of actual history. Worry banished sleep. All night we expected a grim doctor to report that Steve had died. This fear hurt. Hurt horribly. But still there was the possibility he'd recover and start in a clinical trial. That night was awful, but it was not irreversible, as far as we knew. And despite our fears, Steve survived the night, and the next day, and a few more. After about ten days he came home. He got into the clinical trial, and a decade later he's still here.

That horrible night our feelings were not grief. Worry and fear next door to despair, certainly, but not grief.

One of the best-known narratives about grief is *A Grief Observed* by C. S. Lewis.[2] That Lewis projects his grief through the lens of his particular religion simultaneously dilutes and muddies the impression of his reactions, in my opinion, so I won't discuss his book. Your opinion may differ, of course.

I will say something about Joan Didion's books recounting her experiences with grief.[3] *The Year of Magical Thinking* is a subtle, thoughtful account of how she explored her world, irreversibly changed by the death of her husband John Gregory Dunne. Didion focuses on day-to-day events, complicated significantly by the life-threatening illness of her daughter Quintana. I first noticed the complexity of the issues at the end of chapter 2:

> I knew John was dead. . . . Yet I was myself in no way prepared to accept this news as final: there was a level on which I believed that what had happened remained reversible. . . .
>
> I needed to be alone so that he could come back. This was the beginning of my year of magical thinking.

In chapter 17 Didion states explicitly ideas that had percolated through that long year:

> Grief turns out to be a place none of us know until we reach it. . . . We might expect that we will be prostrate, inconsolable, crazy with loss. We do not expect to be literally crazy, cool customers who believe that their husband is about to return and need his shoes.

And she has this to say about anticipatory grief:

> Nor can we know ahead of the fact (and here lies the heart of the difference between grief as we imagine it and grief as it is) the unending absence that follows, the void, the very opposite of meaning, the relentless succession of moments during which we will confront the experience of meaninglessness itself.

Because Didion focuses on details of her life and thoughts, the sophistication and depth of her analysis was a slow burn for me. Her narrative seemed mostly matter-of-fact, until it really didn't.

Didion's *Blue Nights* may be read as a follow-up to *The Year of Magical Thinking*. It is Didion's meditation of the death of her daughter twenty months after that of her husband. Her approach to loss and grief is thoughtful, moving, and presents perceptions I did not expect.

For a visceral example of grief I'll take some passages from the novel *The Dog Stars* by Peter Heller.[4] The narrator Hig and his dog Jasper live in a world where the bulk of the population has died from a new strain of flu. (When I first wrote this section, the COVID-19 pandemic had not begun. I so hope that Heller's novel remains fiction, but in late 2020 I fear that US executive-branch policies that contravene the advice of scientists may push parts of Heller's dystopia into our actual world.) One night Jasper dies. This is Heller's portrait of the grief that consumes Hig:

> In the morning I wake stiff. The sleeping bag and Jasper are covered in frost. So is my wool hat.

> You must be cold, boy. C'mere. I pull his Whoville quilt to fold it over him. He is heavy, unmoving.
>
> C'mon, bud, this'll be better. Til I start a fire. C'mon.
>
> He ignores me. I tug at the quilt and lay it over him, brush his ear. My hand stops. His ear is frozen. I run my hand around to his muzzle, rub his eyes.
>
> Jasper, you alright? Rub and rub. Rub and tug his ruff.
>
> I pull him, stiff and curled, closer to me and lay the quilt over him and lie back. I breathe. I should have noticed. What a hard time he was having on the walk. The tears that weren't there yesterday flood. Break the dam and flood.
>
> Jasper. Little brother. My heart. . . .
>
> Exhausted. To the bone. The untethering that took such an effort. The flight over already seemed like another life. And the airport seemed like a dream. If the airport was a dream, then Jasper was a dream behind a dream, and before before was a dream behind that. Within and within. Dreaming. How we gentle our losses into paler ghosts.

I am in awe of that last sentence, "How we gentle our losses into paler ghosts." This is a picture of grief that puts our minds in Hig's life when Jasper dies. This is not simple sadness. The loss is permanent, the heartbreak is irreversible, though at the end of the quotation Heller notes how grief can become translucent. Eventually it is part of our day-to-day background, another strand in the tapestry of the mind.

If you think that the death of an animal companion cannot sink you into an ocean of grief, I say you've never lost a pet. I won't try to put you behind my eyes or my wife's eyes when one of our cats died, or behind my brother's eyes or his wife's eyes when one of their dogs died. That would fill up the rest of this book. And all those words would be pointless. At any important level, I can never know what you think, nor can you know what I think. Empathy isn't about knowing how another person feels.

Empathy is about how you would feel if you were in the circumstance of another person. This is the best we can do.

Here's an example. When I was ten and my brother Steve was five, one hot summer day Dad and our uncle Bill took us to a sandbar at Lower Falls on the Coal River, not far from St. Albans, West Virginia. That morning several other families were at the sandbar. Steve and I floated in the river for a while, holding an inner tube. Then we got out of the river and looked for interesting rocks by the falls. Dad and Bill swam. A boy about my age waded in the river with his brother, about Steve's age. There was a commotion in the river, and the younger brother disappeared under the water. The older brother yelled for help. Bill threw the inner tube to where the younger kid had vanished. A hand came out of the water, tried to grab the inner tube but slid down the side and was gone. Some of the adults, including Dad and Bill, swam to where the younger brother had vanished and began to dive to the river bottom. The boys' mother wailed. Another adult scrambled up the steep riverbank, found a house, and called the police.

Steve and I sat on the sandbar. The summer sun now felt cold. The older brother sat on the beach not far from us. He pulled his knees up to his chest, wrapped his skinny little boy arms around his skinny little boy legs, and put his head on his knees.

Eventually the police arrived in a patrol boat. They dropped horrible big hooks into the water and began to drag the river. Dad decided it was time for us to leave.

I don't recall much of the conversation with Mom and Dad that night, but they were gentle, thoughtful, and honest. This was the first death I'd seen up close. Ruthie's death was two years in my future. This was the first time I'd seen the incandescent grief of losing a child, of losing a brother. I couldn't imagine how the older brother felt. How could I? I'd need to know so many details of their lives. Were they friends or did they spat often? Did the older brother look out for the younger brother? Did the younger brother add welcome silliness to the older brother's days? To know how the older brother felt, I'd need to know answers to these and a thousand other questions.

The best I could do was wonder how I'd feel if Steve died. If our five years together were all we'd ever have. If he became memory. Stories. And nothing more. Ever. Ten-year-olds shouldn't have these thoughts in the dark hours between midnight and dawn, but I did. I don't know what the brother of the drowned boy felt, but I could try to imagine how I'd feel if Steve had drowned that day. It was a burning, bitter nightmare.

We cannot enter the personal hell of another, but we can imagine our hell if we inhabited their situation. This is how we can think, talk, and write sensibly about grief. Without empathy, grief is trapped within our own heads. Or for some of us whose lives have been heartbreak upon heartbreak, our heads are trapped within grief.

So: grief is irreversible, we cannot grieve for contingent events, there is no anticipatory grief. And whatever shadowy way we understand the grief of another is focused through the lens of empathy.

There are exceptions to my conclusion about anticipatory grief. We can grieve for a friend in the final stages of a terminal disease, or one who has disappeared, for example, in an accident at sea or a military action. You need not stand beside a casket to know someone is gone or soon will be. And while I suppose there's always a chance that a new clinical trial will shock us, or a survivor will be found clinging to a piece of debris in mid-ocean, these are so unlikely that to insist on absolute irreversibility before acknowledging suffering as grief is far too cruel.[5]

Here's a personal example of why. My wife Jean's father, Martin Maatta, died in 1985; her mother, Bunny (Bernice) Maatta, in 2012. We spent a week or two with Bunny almost every summer (winter is not the best time to visit Ishpeming, Michigan). For a decade or so after Martin died, Bunny seemed fine when we visited. We went on day trips around the Upper Peninsula, visited other relatives. In the evenings Jean spread out bank statements and bills across the living room floor, made sure Bunny's accounts were up to date and balanced. Meanwhile, Bunny and I sat up late in the kitchen, swapping stories and jokes. For example, when Bunny was in high school she worked at the Chocolate Shop, the soda fountain in Ishpeming. In the 1930s, many Midwest towns had baseball teams that traveled around in the summer and played evening games. Bunny told me that on days when there was a game in town, she always worked the evening shift. After the game, some of the players would come to the Chocolate Shop for sodas and ice cream. "Those boys were sweaty, muscular, and beautiful," she said. Jean doesn't recall this story; I suppose some things are more easily told to a son-in-law than to a daughter. Bunny and I got along wonderfully. Our visits were a delight.

But eventually we realized that something was wrong. Bunny got confused, repeated herself again and again, and some of the spark was gone from her conversations. Jean hired local health care workers to stay with Bunny, to cook and clean and do laundry, but mostly to keep her company. These were wonderful people; even now, years after Bunny died, we visit them every time we are in Ishpeming. Eventually Bunny's physical needs were more than these women could provide, so Jean moved her mom to a nursing home. There she lived for a decade. Every morning she got into her wheelchair and walked herself up and down the long hallways. By then she mostly didn't recognize us, or anyone, but she was generally happy. The staff loved her. At the end of one of our visits soon after Bunny had moved into the nursing home, Jean wanted to take a photograph of Bunny in her wheelchair. Jean and I backed up a bit from Bunny. She just stared. Several of the staff joined us and waved at Bunny, hoping to get her to smile. Jean asked me to step over to the side and wave at Bunny. I did. She looked at me, broke into a wicked grin, and thumbed her nose at me. Jean had never seen Bunny make that gesture, ever. We all laughed, Bunny too, and Jean got the photo she wanted.

But year after year, less of Bunny's mind was there. Jean and I began to mourn what we thought was the disappearance of her "Bunnyness." Doctors told us that her mind was gone, that her body would follow soon. The emotional weight of the loss was immense and transcendent, and the loss was irreversible. Or so we thought.

On what turned out to be my last visit with Bunny, she'd been very quiet, unengaged. On the last day of the visit, I pushed her wheelchair to the solarium. Bunny held my hand the whole time. The three of us looked at a fish pond and small trees, some birds flying around. Jean stood on Bunny's left, I stood on her right, and she still held my hand. Bunny didn't respond to anything we said. Then she looked at me, frowned, then broke into a beautiful smile and said, "Mike." Then she was gone again. Part of her still was in there, somewhere. Would we talk with her more often, or ever? We didn't know. But we did know that although it was time to be very sad, it was not yet time to grieve.

With death the cause of grief is obvious. Other circumstances—grave illness or dementia, for example—are more complicated. How close we are to what I call grief depends on how certain we are that the loss is irreversible. At the moment this is shifting ground. You'll need to sort out this balance on your own, if you try this approach.

* * *

Now I'll sketch a bit of the history of the study of grief, using psychologist John Archer's 1999 book *The Nature of Grief* as my main source.[6]

In the 1970s psychologist John Bowlby argued that grief,

which is maladaptive, is a separation reaction.[7] Generally, separation reactions are adaptive; that is, they cause us to seek the loved individual from whom we are separated. The circumstances where such a reaction is useful greatly outnumber those where it is not, so evolution has not selected against it. But grief is a separation reaction in a circumstance where reunion is impossible.

Psychiatrist Colin Parkes proposes that grief is a necessary consequence of the way we form attachments.[8] Bowlby and Parkes explain that the separation anxiety response has evolved genetically for associations important to the individual's survival (the young child depends on the parent for protection and food) and to send the individual's genes into the future (the parent depends on children for this). Archer speculates that the recognition of death arose only recently in human evolution, and we have not yet had time to develop distinct responses for separation with the possibility of return and for irreversible separation, such as death.

Archer reports on studies that sought to reduce the variety of grief reactions to a single variable. (Really? Who thought this was a good idea?) Using a statistical technique called factor analysis (a very interesting method, next-door to magic; if you aren't afraid of statistics and don't mind a bit of math, you should look it up), researchers tried to group together various modalities of grieving. The results revealed little consistency. I read this as saying that grief is inherently high-dimensional.

We build complex mental models of loved ones, and these models are incorporated as parts of our self-model. When the environment signals a mismatch, for example, when the loved one no longer can be found, separation anxiety is triggered. We try to align environmental signals with model expectations. When this can't be done, we must adjust our self-model. This takes time and effort.

The first detailed psychological study of grief is incorporated in the *laws of sorrow* of Alexander Shand.[9] Because experimental data was lacking, Shand turned to literature and poetry to illustrate his ideas. Even when experimental data are available, the arts can provide glimpses of raw emotion more directly than any number of psychological studies.

A commonly held view is that grief passes through several stages or phases.[10] In fact, Archer writes, "the phase view clearly lacks empirical support," but it "does represent an attempt to capture the dynamic nature of the process of grief." A version of these stages figure in "One Fish, Two Fish, Blowfish, Blue Fish," season 2, episode 11, of *The Simpsons*. At a Japanese restaurant Homer eats sushi contaminated with blowfish toxin. At the hospital Dr. Hibbert examines Homer and announces that he'll die within a day:

HIBBERT: You can expect to go through five stages. The first is denial.
HOMER: No way, because I'm not dying.
HIBBERT: The second is anger.
HOMER: Why you little. Grrr!
HIBBERT: After that comes fear.
HOMER: What's after fear? What's after fear?
HIBBERT: Bargaining.
HOMER: Doc, you gotta get me out of this. I'll make it worth your while.
HIBBERT: Finally, acceptance.
HOMER: Well, we all gotta go sometime.
HIBBERT: Mr. Simpson, your progress astounds me.

To be sure, this does not reflect the experiences of most people. But I hope you'll agree that the interchange is funny.

Implication can lend import to names. "Recovery" from grief

suggests a return to the circumstance before the loss, which is impossible. The dead person still is dead. "Readjustment" signifies an adaptation of our world model to include the absence, in which we modify all aspects of the model that touch the person who has gone. Life can continue, but it will not, it cannot, be as before. I cannot see a biscuit without memories of my mother baking them for special breakfasts. The warm kitchen filled with the aroma of biscuits. Selecting the jams, setting the table, were painful memories for a year or so after Mom died. Now these same memories bring sadness but also fondness, an ever deeper appreciation of how very lucky I am to be Mary Arrowood's child. Biscuits are for me what madeleines were for Marcel Proust. This is not recovery; it is readjustment.

One traditional approach to dealing with bereavement, called *grief work*, has four components: accept that the loss is real, work to reduce the pain of the loss, modify your identity to accommodate the loss, and detach yourself emotionally from the person who has died. Though popular at one time, the grief work hypothesis now has some opposition.[11]

The *dual process model* (DPM) developed by Margaret Stroebe and Henk Schut postulates two processes: *loss-oriented*, which involves confronting grief, and *restoration-oriented*, which involves attending to other aspects of life.[12] Readjustment involves "oscillation" between these two processes. This can aid in revising our models of world and self. For example, establishing a new relationship to replace the one broken by death should not advance the resolution of grief according to the grief work hypothesis, but should according to the dual process model. Can remarrying help a widow or widower?

Does grief have an evolutionary basis? In *The Nature of Grief*, Archer writes that grief is a consequence of our need for relationships, parent and child being the central one, though social animals (including us) form other important attachments. Many of these have survival value, so broken attachments may pose survival risks. Separating a child from its parents puts the child at risk by removing parental protection, and also threatens the parents' effort to send their genetic material into the future. An important step in this attachment theory is the notion of kin selection advanced by evolutionary biologist William D. Hamilton's (not to be confused with the mathematician William R. Hamilton). Kin selection is an evolutionary process that favors the reproductive success of relatives, even at some expense to one's own reproductive success.[13] Hamilton quantifies a relation that makes kin selection viable:

> (the degree of relatedness) times (the benefit to the relative) must exceed the cost to the individual.

Both attachment and separation thus have symmetrical aspects. Mutation discovered, and natural selection amplified, stress hormones that motivate efforts to avoid the dangers of separation.

This is not just true for people: animals, too, can experience separation anxiety and can grieve. Patch, named for a gray patch of fur around her right eye, is a feral cat who lives mostly in our yard. Several years ago she had a litter of four kittens in a house under construction near our house. The workers brought the kittens to us in a box, and we took them to a no-kill shelter in a neighboring town. As near as we could tell, the four shelter employees on duty when we delivered the kittens each took one home. The kittens were impossibly cute. For the next week, Patch wandered around our yard, crying for her kittens. She certainly sounded forlorn; she ate little or nothing, just looked for her kittens. This appeared to be a clear example of grief in response to separation. Eventually she stopped searching and began to eat more regularly.

Patch often spends time around Slinky, another feral cat, named for how he slinks away when someone approaches him and not, as my brother suggested, for how he goes down stairs. We think Patch and Slinky are littermates. Jean and I trapped them one at a time to be neutered and vaccinated. When each was trapped, the other paced around the cage and cried. When we brought each home from the vet and released it, the other appeared. They rubbed their heads together and walked off side by side. Even now, years later, they accompany one another. In fact, as I write this, from my workroom window I see them curled up together in our backyard.

I wonder if for the day Slinky was gone Patch feared he would not return? To me, her reaction seemed frantic, similar to that when she lost her kittens, but I cannot know what is in a cat's mind. Both looked like grief to me. But can a cat understand irreversibility? Are my criteria for grief relevant across species?

Back to the evolutionary basis for grief. Separation activates stress hormones. Stress itself has no survival value, but the effort to alleviate the stress—to find the missing partner—does have survival value if it succeeds. But in circumstances when the partner has died and so cannot be found, grief, with all its negative consequences, seems to have no positive survival value. In fact, Archer says that, rather than having any survival value, grief is an epiphenomenon of our capacity for attachment, that is, a secondary consequence of selection for the mechanics of attachment.

An evolutionary explanation of grief, then, is the stress hormones of separation coupled with the realization that the other person will never *never* NEVER return. Writing about this is very difficult. Time folds up, so many ghosts crowd into my head. Parents, grandparents, aunts and uncles, dear friends, students. (How can the student die before the teacher? Something is seriously wrong with the universe when this happens.

Adam, we had so many more projects to do. Why the hell did you smoke cigarettes? I wonder what else your programming and my math would have found.) And far too many cats. Never again will sweet little Bopper curl up above my head on the pillow, his purrs my lullaby.

We can push a bit further in this direction with Barbara King's book *How Animals Grieve*, though I have to say that for me this book was a challenge to read.[14] For people who love animals, the subject is emotionally demanding. King is a compelling writer, and she presents most of the material through stories about particular animals. This approach does help us understand that animals can love, that animals can grieve.

We may not be surprised that chimpanzees and elephants

grieve; after all, both elephants and chimpanzees use tools and engage in play, thus demonstrating considerable cognitive sophistication. And our own experiences with our dog and cat companions provide ample evidence that dogs and cats can feel grief. But dolphins and whales? Turtles? Chickens, rabbits, pigs, ducks, geese, monkeys, storks, crows and ravens, horses and goats and buffalos? Yes to all.

Animal grief is predicated on animal love. Based on the work of Jane Goodall, Cynthia Moss, Marc Bekoff, Peter Fashing, and others, and on her own fieldwork in Kenya, King offers this formulation of animal love:

> When an animal feels love for another, she will go out of her way to be near to, and positively interact with, the loved one, for reasons that may include but also go beyond such survival-based purposes as foraging, predator defense, mating, and reproduction.

and

> Should the animals no longer be able to spend time together—the death of one partner being one possible reason—the animal who loves will suffer in some visible way. She may refuse to eat, lose weight, become ill, act out, grow listless, or exhibit body language that displays sadness or depressions.

So King includes the potential for grief as a component of love. In fact, she calls this a sufficient condition for love. But what about the irreversibility that I think is an essential component of grief? King writes that "animal grief does not depend upon a cognitive mastering of the concept of death." Through example after example, she supports the notion that animals can intuit the permanence of loss that follows from death.

King recounts many instances of cats who, when a sibling or partner is lost, wander in search of the lost partner. Eventually, the cat's search is accompanied by howls, by shrieks, by keening. It is difficult to understand what this is, if it is not grief.

We know that people grieve in different ways, perhaps not at all visible to others. With this in mind, King encourages latitude in the interpretation of what we observe: "We shouldn't require every dog to grieve in order to believe that some dogs do."

Animals grieve in different ways. Horses form a "horse circle" around the grave of a fallen member of their herd. Cows have exhibited similar behaviors. And the elephants: a dead elephant is visited and apparently mourned not just by blood relatives but by elephants from other families.

There is an old view that all non-human animals are stuck in time, that is, they have no sense of past or future and so the perception of irreversibility is not open to them. But mounting evidence suggests that many animals exhibit episodic memory (particular memories of an event, unique to the individual) and perhaps autobiographical memory (memory of an personal history).[15] Then there is King's description of the chimpanzee Brutus, who exhibited double-anticipation when hunting: he anticipated the movements of both his prey and the other chimps in the hunting troupe.[16] King deduced that Brutus "could reflect on the mental state of others," that he has a theory of mind.

The notion that human minds have access to a wide variety of modalities closed to animals seems supported more by arrogance than by evidence. Maybe we should not be so hasty to believe that anticipation of the future is a purely human trait. Nor is the perception of irreversibility, though I am not yet sure where in this analysis to place King's reports that some female monkeys carry dead infants for days.[17] Do the mothers hope their babies may return to life? Is this their way to mourn? We have way more questions than answers.

The range of animal responses to death is immense. King's book will open your eyes, but have a box of tissues nearby when you read *How Animals Grieve*.

Focused not on grief, but on understanding the world as it appears to a member of a different species, a goshawk named Mabel, is Helen Macdonald's wonderful *H Is for Hawk*.[18] I was particularly beguiled by the descriptions of how Helen and Mabel played. Cats and dogs play, and I've seen squirrels act in ways that I think of as play, but I hadn't known that birds can play. And if birds do play, I'd look first to sparrows or wrens or finches, not scary raptors. This was a revelation.

To see the world as another sees it, we might have more success with another species than with another person. What another person sees always is filtered through what we see, projected onto our categories. To see what a hawk sees, we must erase all that. Start with a clean slate and through long, careful observation (this is the hard part, hundreds or thousands of hours of close company, of noting reactions to shared experiences) build up a glimpse of a world mostly unseen by us. Consider this:

> I become both the hawk in the branches above and the human below. The strangeness of this splitting makes me feel I am walking under myself, and sometimes away from myself. Then for a moment everything becomes dotted lines, and the hawk, the pheasant and I merely elements in a trigonometry exercise, each of us labelled with soft italic letters.

H Is for Hawk is Macdonald's story of living with and training a goshawk; it also is a story of her grieving for her dead father. Not surprisingly, she approaches this from a different perspective.

> Ever since my father died I'd had these bouts of derealisation, strange episodes where the world became unrecognizable.

(In chapter 4 I'll mention a tiny echo of this experience that my brother and I shared. The word "hound" is the bookmark.)

And then there's this:

> The archaeology of grief is not ordered. It is more like earth under a spade, turning up things you had forgotten. Surprising things come to light: not simple memories, but states of mind, emotions, older ways of seeing the world.

If you haven't read this book yet, you should read it soon. You'll learn things sad and also things beautiful.

Physician and scientist Randolph Nesse has written a penetrating analysis, "An evolutionary framework for understanding grief."[19] With the evolutionary biologist George Williams, Nesse wrote the brilliant book *Why We Get Sick: The New Science of Darwinian Medicine*.[20] They investigate disease through the lens of evolution; their conclusions are eye-opening. One short example: typically, a low-grade fever does little harm beyond burning a bit more resources. But raising body temperature even a couple of degrees can impede significantly the growth of pathogens and allow the adaptive immune system time to identify the attacker and amplify the supply of appropriate antibodies. Taking aspirin to lower a mild fever can be exactly the wrong thing to do. You should read this fascinating book as well. You see that I'm not shy about reading recommendations.

So, Nesse is perfectly positioned to provide a subtle evolutionary analysis of grief. The focus of his study is this question: "What selection forces shaped the brain mechanisms that give rise to grief?" Natural selection discovered the processes that produce emotions by adjusting physiology. For example, when our Pleistocene ancestors saw a distant predator, they became anxious, and that anxiety helped them avoid the predator. Negative emotions can have a high cost, so unless they also have some survival value, negative emotions experienced up through the reproductive years would have been removed by natural selection. Sadness occurs after a loss and can help us engage in several responses: trying to reverse the loss, acting to prevent future losses, warning others about an ongoing danger. When the loss is irreversible, when the sadness free-falls into grief, the cost of this emotion does not help us reproduce. But the actions we take to combat similar losses may increase the likelihood of our children's survival.

I must mention that Nesse asks, "Is grief a special kind of sadness shaped to cope with the adaptive challenges posed by loss of a close relative or a loved one?" He believes that evidence supports this. But mostly this is his approach to counter Archer's epiphenomenon interpretation. Nesse's category of sadness appears to be more inclusive than mine.

How people experience and express grief, and how they readjust their lives to accommodate loss, is among the most personal aspects of life. This is no surprise: grief is tied to love and love is *the* most personal of experiences. A corollary is that what happens to us after a loss varies from person to person.

My friend the playwright Andrea Sloan Pink has described some of her experiences after she lost her mother. Everyone endures loss and grief inside their own minds. We hear what the grieving person says, we try to understand the pain in their world, but we will fail. Let them speak, but don't offer words to comfort. If you're in a position to help with their day-to-day, offer that. This is why friends bring food to a bereaved family. Otherwise, listening is the best we can do. Listen to Andrea now:

> Numbing and scalding. My mother's death gave two long physical sensations, burning beneath the skin, and a terrible internal bright light, as bad as an eye exam but worse. Eventually, these strange neurological sensations relax, but I'm not

entirely sure that's good. I didn't want people to convince me not to make changes, to be "satisfied" again with that which was not satisfactory.

One of the things that shocked me about my mother's death was the universe's high tolerance for waste. How can the universe afford to take a consciousness that can speak five languages one day and dissolve it the next? All the utility and investment is squandered.

Grief was like a blinding light that shone over my life to make stark its inadequacies. So many things felt wrong. This has eased. But I remember some of those lessons that I saw in high contrast.

3
Beauty

Decorations all in royal blue

Beauty is a bridge between grief and geometry. Demonstrating this will take some work.

Barbara King presents evidence that grief is tied to love. She formulates this for animals:

> Grief blooms because two animals bond, they care, maybe they even love—because of a heart's certainty that another's presence is as necessary as air.[1]

That this applies to human grief needs no support beyond our own experience. In this chapter we'll tie grief to another strong emotion, our response to beauty.

Already we've linked geometry to beauty: some parts of geometry are so beautiful they take my breath away. Now I'll argue that beauty and grief are next-door neighbors, or maybe grief is beauty in a dark mirror.

In one sense this is obvious: both beauty and grief constrict the airways, or maybe paralyze the diaphragm. Any intense emotion can impede our breath, so gasps alone do not provide a tie between grief and beauty. For this we must look a bit more deeply.

We've talked about grief already and have found some characteristics, so to look for a tie between beauty and grief, we must unpack some properties of beauty. Just as we distinguished grief from sadness, we must separate beauty from prettiness. I'll start with a memory from my childhood and hope it leads you to your own parallel memories.

Some evenings shortly before Christmas, we piled into the family car and Dad drove us through neighborhoods in St. Albans to look at decorations. Many trees were lit in multicolored explosions of red, green, blue, and yellow. These Mom called "pretty." Some others were monochrome, all blue lights or all white lights—"beautiful," according to Mom. I was a curious little kid, so surely I asked her to explain how to tell "beautiful" from "pretty." Sadly, I don't recall her answer, hardly a surprise because this happened in the late 1950s. Mom died years ago so I can't ask her; I'll try to figure out what she might have said.

In Western philosophy the roots of the study of beauty extend to the ancient Greeks, and perhaps further.[2] Our simple analysis doesn't need a complete theory of aesthetics. Rather, I'll mention three authors—Daniel Berlyne, Denis Dutton, and Richard Prum—to guide us.

In his book *Aesthetics and Psychobiology*, the experimental psychologist Daniel Berlyne writes that in order to be perceived as aesthetically pleasing—surely a requirement for both beauty and prettiness—two characteristics are required: novelty and familiarity.[3] Novelty provides the element of surprise. The sound of someone practicing musical scales again and again exhibits no novelty; it is not interesting, not aesthetically pleasing. Familiarity, on the other hand, is needed to provide a context—if not a path to understanding the composer's intent, at least a map to fit the piece into our experience. Radio static is not aesthetically pleasing because it has no recognizable pattern, nothing familiar. So both beauty and prettiness must exhibit aspects that are novel as well as aspects that are familiar.

This theme has its roots in Berlyne's paper "A Theory of Human Curiosity," based on his Yale doctoral dissertation. Berlyne concludes that patterns will be most curiosity-arousing if their familiarity is at an intermediate level. His analysis of curiosity is based on the notion of conflict between possible responses, where the degree of curiosity is correlated with the degree of this conflict. Patterns too unfamiliar will not produce responses adequate to generate much conflict, while patterns too familiar do not generate conflict because the pattern is expected. Curiosity arises most strongly in the Goldilocks zone between novelty and familiarity.

Berlyne's balance of familiarity and novelty can be viewed as an extension of George Santayana's notion that the experience of beauty follows from a delicate balance of purity and variety.[4] Santayana formulates this in his analysis of the sense of beauty of sound, but he states that his analysis is "a clear instance of a conflict of principles which appears everywhere in aesthetics." Here's Santayana's version of Berlyne's balance:

> Since a note is heard when a set of regular vibrations can be discriminated in the chaos of sound, it appears that the perception and value of this artistic element depends on abstraction, on the omission from the field of attention, of all the elements which do not conform to a simple law. This may be called the principle of purity. But if it were the only principle at work, there would be no music more beautiful than the tone of a tuning fork. . . . The principle of purity must make some compromise with another principle, which we may call that of interest. The object must have enough variety and expression to hold our attention for a while, and to stir our nature widely.

Berlyne doesn't mention Santayana in *Aesthetics and Psychobiology*. This suggests that the notion of novelty and familiarity—or variety and purity—as necessary for aesthetic appreciation, was naturalized by the time he wrote his book. It was simply "in the air."

Next, in his book *The Art Instinct: Beauty, Pleasure, and Human Evolution*, philosopher Denis Dutton presents a Darwinian theory of beauty.[5] Dutton pushes back against the common doctrine that aesthetic taste is culturally conditioned. Even a moment's thought shows that our sense of beauty is not restricted to our own culture. Did you think the aerial battles in the bamboo grove of Ang Lee's *Crouching Tiger, Hidden Dragon* were beautiful?[6] The ballet of Li Mu Bai and Jen Yu, graceful and fraught; the slow sway of the bamboo against flashing swords; the rumble of distant thunder and Yo-Yo Ma's cello.

Did the final pages of José Saramago's *Death with Interruptions* leave you gasping?[7] With Saramago the plot is never, ever the point; his imagination, his heart, and especially his language are. The novel opens with the sentence "The following day, no one died." This lack of death goes on for some time, and while you might think this a blessing, it is not. People still get sick and injured; they just don't die. Many complications arise. Eventually, death decides to get back to work, but accustomed now to not dying, people complain about not having time to prepare. So death starts sending letters, in violet envelopes, that tell people they will die in exactly seven days. This does not please people, either. One day, one of death's letters is returned to her. She sends it again, and again it is returned. So she takes human form to find the person whose letters have been returned. He is a cellist. She gets to know him, eventually falls in love with him, and they sleep together in his apartment. Then

> He fell asleep, she did not. Then she, death, got up, opened the bag she had left in the music room and took out the violet-colored letter. She looked around for a place where she could leave it, on

> the piano, between the strings of the cello, or else in the bedroom itself, under the pillow on which the man's head was resting. She did none of these things. She went into the kitchen, lit a match, a humble match, she who could make the paper vanish with a single glance and reduce it to an impalpable dust, she who could set fire to it with the mere touch of her fingers, and yet it was a simple match, an ordinary match, and everyday match, that set light to death's letter, the letter that only death could destroy. No ashes remained. Death went back to bed, put her arms around the man and, without understanding what was happening to her, she who never slept felt sleep gently closing her eyelids. The following day, no one died.

The repeated descriptions of the match, unnecessary for the plot, knocked me down the first time I read this passage. While successive readings don't match that intensity, they do remind me of Saramago's genius. For those who say the novel is just an extension of an early 1960s *Twilight Zone* episode, I have only sympathy for your lack of imagination.

What about Mbuti barkcloth paintings, or Inuit animal sculpture, or the Lascaux cave paintings, or the mosque in Cordoba (sketched on the next page)?[8] There are many examples. Art of every culture can be appreciated by people in every culture, so if the appreciation of art is culturally conditioned, that conditioning is indeed subtle.

My father was not well-educated. He quit high school to work in the Newport News shipyards until he turned seventeen and joined the Navy in World War II. He was a skilled carpenter and millwright but had little exposure to art. When an image of a modernist painting flashed across the television screen, Dad made the familiar pronouncement, "A five-year-old kid with a box of finger paints could do better than that." I showed him images from a book of Paul Klee paintings.[9] He looked at them,

quietly, for a while, whistled, and said, "Well, no, I don't suppose a five-year-old with finger paints could do those." Then he asked a question that points to a central issue of how to understand beauty: "How can a picture look so good and not be a picture of something?" The image he was looking at was not of any object or scene he recognized, but still it drew him in. Dad grew up in Rosedale, West Virginia, during the Great Depression. He did not enjoy school, though he told me he liked math, but that may have been just a kindness to me. He served in the Pacific during World War II. After the war eventually he worked for Union Carbide, initially as a day laborer, then in the powerhouse, then in the pump house, and finally as a millwright in the machine shop, his dream job. Along the way he married my mom and they raised three kids. I doubt he ever set foot in an art museum or gallery. And yet, Klee's paintings moved him in ways he could not understand. Yes, the perception of beauty is subtle.

A Darwinian approach to anything is based on selection, but Darwin put forward *two* selection principles. The more familiar is *natural selection*, presented in *On the Origin of Species*.[10] Any trait that improves the likelihood of survival through the reproductive years is more likely to be passed on to offspring, so the fraction of the population with this trait is amplified. Random mutations drive the exploration of many traits; natural selection weeds out those that are harmful. This is such a simple, elegant idea. Little wonder that people who understand it are smitten.

The second principle is *sexual selection*, proposed by Darwin in his second book *The Descent of Man*. Roughly, sexual selection is the idea that females select mates with some traits just because they find these traits aesthetically appealing. Dutton's application of Darwinism to the perception of beauty works to explain why art can be appreciated across cultures. "Beauty is nature's way of acting at a distance," Dutton says. That is, the appreciation of beauty derives pleasure from looking at something rather than from eating it, clearly the better choice if "it" refers to your mate or your child. From our deep history Dutton points out that stone hand axes, produced in abundance by *Homo erectus*, too numerous and mostly unused, were not likely made as tools for butchering animals. Dutton thinks they are very early art, "contemplated both for their elegant shape and their virtuoso craftsmanship." Then sexual selection kicks in because craftsmanship signals skills desirable in a mate.

Finally, ornithologist and evolutionary biologist Richard Prum reverses Dutton's direction. In *The Evolution of Beauty* Prum points out the early opposition to Darwin's idea of sexual selection by mate choice: the idea of female agency in mate selection was too feminist for Victorian England.[11] Alfred Wallace, a vocal supporter of Darwin's evolution through natural selection, was an equally vocal critic of sexual selection. Wallace insisted that natural selection explains everything.

The story of the reception of aesthetic selection is complex, so we'll sketch a few points. Prum has assembled four decades of field observations of bird mating rituals to support the role of sexual selection. In a lecture on aesthetic selection to prospective science students at Yale, Prum showed videos of some rituals. Particularly amusing—I suppose "interesting" is a more neutral word—was the dance of the male superb bird-of-paradise. (You should look it up on YouTube; I can't imagine an adequate verbal description.) In the question period after the lecture, the astrophysicist Meg Urry, a discussion leader in the course, asked why, after seeing that display, the female superb bird-of-paradise doesn't laugh so hard she falls off the branch. A convincing answer was not given.

In 1915 statistician and geneticist Ronald Fisher explained the evolution of sexual ornament with the observation that ornament should evolve to match the mean preference.[12] But then, how does preference evolve? Fisher proposed a two-stage model. Initially, the rudimentary ornament does indicate robust health or some other trait with real survival value. Once preference (sexual selection) based on this ornament is established (the tail of a male peacock is a familiar example), the ornament can disconnect from its correlation with survival value and be selected just because potential mates find it attractive.

Following and extending Wallace's ideas, in 1975 evolutionary biologist Amotz Zahavi introduced the *handicap principle*, that ornamentation is a survival handicap, the existence of which demonstrates that the ornamented individual must have superior traits because it has survived in spite of this handicap.[13] Many biologists found this argument convincing. Many still do.

In 1986 evolutionary biologist Mark Kirkpatrick proved that if the survival disadvantage of an ornament is directly proportional to the sexual advantage of the ornament, then evolution will favor neither the ornament nor the mating preference.[14] Consequently, selection will not amplify the trait in the population. Then in 1990 ethologist and evolutionary biologist Alan Grafen showed that if the relation between ornament disadvantage and mating preference is nonlinear, then the handicap principle could explain the evolution of ornament.[15] Note: "could," not "must." The debate continues.

Prum interprets the accumulated evidence to support the notion that ornament and preferences coevolve outside the working of natural selection, that is, that both aesthetic selection and natural selection drive evolution. Aesthetic selection generates variations that are historically contingent; they depend in detail on the sequence of events and can go in directions that are, well, quirky. Prum's formulation that "beauty happens" supports a greater diversity of forms, which appears consistent with observed variation among the ten thousand bird species. The complexity of these issues illuminates the vigor of the investigation into the role of aesthetic selection.

So: beauty has features familiar and unfamiliar, and evolution and beauty are tied up in very complicated ways. This evolutionary aspect will help us to see relations between beauty and grief.

First, evolution can provide some insight into the distinction between the pretty and the beautiful. Remember the Christmas light story. Pretty is what we see; beauty hints at something deeper, transcendent. Because trees and lights are from such different worlds, combining them already suggests novelty. (Note that a combination of disparate objects does not necessarily imply beauty. No one would think that Christmas lights on a bowl of corn flakes is beautiful.) Familiarity comes from the fact that both categories of objects are parts of our everyday world. Perhaps Berlyne's theory of curiosity explains the smaller emotional response to multicolored lights: "patterns too unfamiliar [a swarm of multicolored lights] will not produce responses adequate to generate much conflict." Possibly, but instead we'll follow Prum's study of the bowers of male bowerbirds. The bower is two parallel walls of twigs (Google an image—you'll be impressed) built to entice a female. As Prum explains, "The male Satin Bowerbird gathers objects with which to decorate it, all of them royal blue, and he piles them on a bed of straw in the courtyard area located at the front of the bower. . . . In most Great Bowerbird populations, males collect and display light-colored pebbles, bones, and snail shells for their bowers."[16] Aesthetic selection has discovered that an arrangement of branches decorated with monochrome objects is beautiful.

Almost surely Mom did not explain her aesthetic choices in relation to bowerbird courtship rituals. What would she have thought if her ten-year-old son had mentioned bowerbird courtship? I expect she said that the pretty lights were too "busy" to be beautiful. Beauty was somehow purer, or simpler. The world in all its complicated messiness can be pretty. Mom wouldn't have used the word "transcendent," though certainly she knew that word because she read a lot. But I believe this was the feature she tried to describe, her notion of what separates beautiful from pretty.

Our branch of the phylogenetic tree separated from birds' branch over three hundred million years ago. Is it reasonable that this monochromatic sense of beauty has been encoded genetically in us and in birds from our common ancestor? While the function of much of our genetic code remains unknown, I don't think this notion of inheritance of a sense of beauty is plausible. More likely this sense has arisen, separately and independently, along several evolutionary paths. If you think this is unlikely, remember that eyes have evolved, separately and independently, perhaps forty times throughout the history of life on earth.

Have other species developed appreciation for monochromatic beauty? Many, though certainly not all, flowers have only one color. Many, though certainly not all, birds have only a few colors and their plumage is dominated by a single color. For example, many swans are mostly white, and many male cardinals are mostly red. Sexual selection seems to have discovered this pattern many times. Why? In an abstract perceptual space this seems to be situated on a local peak, to borrow geneticist Sewall Wright's notion of fitness landscape.[17] For many species, the notion of beauty occupies a position of maximum adaptation—at the moment. Nothing is fixed because each species evolves against a background of the evolution of other species. More completely described, evolution is coevolution: we're all in this together.

The transcendence of beauty is the last piece we need to see the connections between grief and beauty, and between beauty and geometry. Our experience of both grief and beauty involves perceptions of great emotional weight that irreversibly change our circumstance. In addition, our experience of both grief and beauty involves transcendence. To see beauty is to glimpse something deeper; to grieve is to glimpse a loss whose consequences we will not unpack for years, and maybe never.

The beauty of geometry likewise involves great emotional weight, irreversibly alters our perceptions, and is transcendent. For we don't see all of geometry, only a hint, a shadow of something much deeper. Our thoughts about beauty are the mirror we need to see common features of grief and geometry.

We've had enough general arguments for a while. These connections I'll present through stories. Not because I think mine are important, but rather because sometimes subtle ideas can be communicated more substantively through stories than through generalities. Also, I hope my stories will bring to mind your stories. That the landscape of your internal world should agree with mine seems unlikely. I hope that your memories lead you to conclusions that differ from mine. But if we begin to understand the differences in how we model the world, we can each refine our own models.

My tenth-grade geometry class was wonderful. The bits of geometry I'd seen in other math classes and books were assembled all together in one place. The organization and flow of a proof is beautiful, crystalline, pure. (Okay, I admit that not all of my classmates saw proofs as delightful; for some kids, proofs were boring slogs. To them I say, "Your loss.") And then there were the questions. The ancient Greeks defined the number π to be the ratio of circumference to diameter of a circle, but why should this ratio be the same for all circles? There is an answer, and it is fairly elegant, but to a curious tenth grader, this was a delicious mystery. At home at my desk in the evening, at work on the day's assignment or some extra reading, sometimes I'd look out the window. As the evening sky deepened, purple to indigo to black, and a few stars appeared, I wondered if some of those stars had planets that were home to creatures who could contemplate their environments. If they abstracted the shapes of their world into a geometry, I thought it would be the geometry I knew. This sense of universality was amazing.

A lot of thought, and some very interesting conversations with my teacher, led to one explanation: geometry encodes facts about the structure of space and time. The ideas fit together so cleanly, so perfectly. When a proof was completely in my mind, when I saw how each step worked and why, I had my first small taste of a joy subtle beyond all common measure.

Then there was my geometry teacher, Mr. Griffith. Youngish, balding, and absolutely transparently in love with geometry. For me, this love was contagious. Do you see why I fell in love with geometry when I was fifteen? At age sixty-nine I remain as in love now as I was then.

Teachers weren't paid very well then, a situation that, outrageously, has not improved all that much. Mr. Griffith had a part-time evening job as a computer operator for the West Virginia State Road Commission. He invited me to visit the computer center, and one evening my grandfather drove me to the SRC offices in Charleston. I remember a large room full of computers the size of refrigerators, reel-to-reel tape drives, panels with flashing lights. Mr. Griffith explained what the pieces of equipment were and what they did. He described the problem that occupied the computer, a simulation of traffic flow on the West Virginia Turnpike. Math in action, solving a real problem, right now. I'd known about this sort of work. After all, astronaut John Glenn's orbital spaceflight had occurred when I was in elementary school. (Unknown to me at the time, mathematician Katherine Johnson, who did many of the launch and landing calculations for NASA, years earlier had gone to high school across the Kanawha River from my home.[18]) Still, this was tangible. I could see these machines, touch them if I wished. Math had become a physical thing.

Near the end of the spring semester Mr. Griffith and I had a discussion about the emotional impact of learning geometry. By then we'd seen more complex techniques. The proofs were longer, more subtle, and by any measure I could imagine, more beautiful. But they weren't as much fun as those from the beginning of the fall semester. We talked about several possible reasons, including that longer proofs are more difficult to fit into your head all at once. But then Mr. Griffith turned the conversation sideways and asked what was my favorite piece of music. Bach's fifth Brandenburg Concerto was the first that came to my mind. How many times had I listened to it? Dozens, at least. Did I remember the first time I'd heard it? Yes, clearly, at the house of my friend Gary Winter. How did I feel about it? I'd never heard anything like it, chills down my spine, the patterns were so beautiful. Did I feel that way when I listen to it now? Not exactly: I hear more variations in the patterns, but the baseball bat between the eyes of the first time I heard it hasn't repeated.

"That's the trouble," Mr. Griffith said. "The first time you hear or see something beautiful can be the most intense. Sometimes it seems like that feeling for the thing dies at the end of the first experience. You only get one chance to see a proof of the Pythagorean theorem for the first time."

This idea has haunted my thoughts for years, reinforced when I began to learn logic, coding, quantum mechanics, general relativity, differential topology, fractal geometry, dynamical systems, and most recently, mathematical biology. All of these involve "firsts." For instance, when you first learn about Gödel numbering, you assign numbers to variables and to logical operations, then to propositions.[19] If you do this carefully (and Gödel was very careful), you can write propositions that refer to their own numbers. This allows the self-reference that Gödel used to prove the incompleteness theorem. The sheer brilliance of this idea, the surprise that it will work, leaves room for only one response, "a tingling in the spine, a catch in the voice, a faint sensation, as if a distant memory, of falling from a height."[20] Reviewing the proof, studying it again and again, can reveal nuance missed in your initial exploration, but it cannot reproduce the absolute sense of awe in the presence of such beauty.

When I see something beautiful, that first realization is tinged with grief, because I know I'll never again feel so strongly about it. When I see something pretty, there is no initial gasp like the gasp that accompanies a first glimpse of beauty. Subsequent viewings of the same pretty thing can produce about the same pleasure. We feel no grief, because our initial impression is reproducible.

Part of the grief of geometry arises from this: our first view of a beautiful geometric construction realigns our thoughts in a way that cannot be reversed. We cannot, for a second time, have a first view.

I'll give another example of this dimension of grief, returning one more time to fractal geometry. For the twenty years I taught this course at Yale, the first day was a survey of the basic notion of self-similarity. Recall from chapter 1 that the Sierpinski gasket consists of three pieces—in the figure on the next page, lower left, lower right, and upper left. Each is similar to the whole shape, hence the description self-similar. I would follow with a bunch of natural examples: ferns, trees, river basins, coastlines, mountain ranges, clouds on Earth, clouds on Jupiter, clouds of stars, our lungs, our circulatory and nervous systems, some poems by Wallace Stevens, many (sufficiently long) pieces of music, and more and more. The theme of self-similarity reveals a symmetry—symmetry under magnification—that gives another way to understand many shapes in nature.

The second class focused on finding simple rules that generate fractal images. We start again with the Sierpinski gasket. Now shrink the entire gasket by one-half and send the shrunken

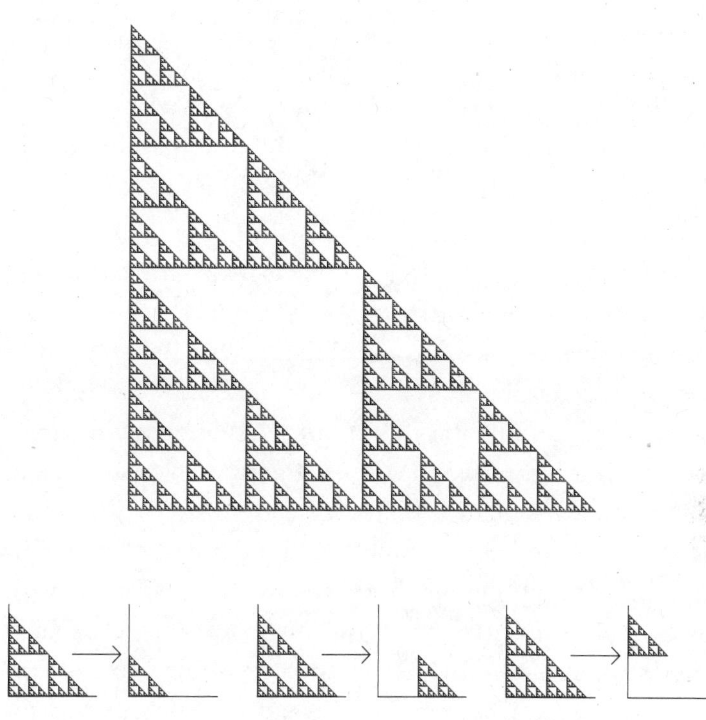

gasket to the position of the lower left-hand piece of the original gasket. Again shrink the whole gasket by one-half and move it to the right half the length of the gasket base, to the place of the lower right piece (middle image). And a third time, shrink the whole gasket by one-half and move it up half the length of the gasket side, to the place of the upper left piece (right image). Apply these three rules to the gasket and you get the gasket. In fact, the gasket is the *only* shape left unchanged by these three rules.[21] Apply them to any other shape and you don't get that same shape. Try, for example, applying the three gasket rules to a sketch of a cat (see the next page). After one iteration, we get three smaller cats. And suppose we apply the three rules to these three smaller cats. We get nine still smaller cats. Continue and eventually the family of cats appears to turn into the gasket.

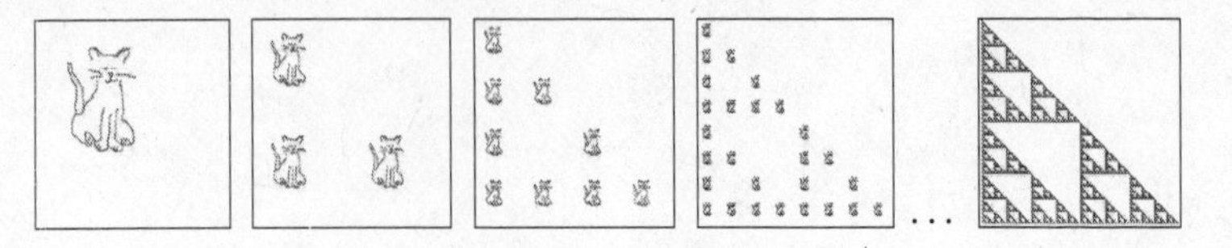

Now after any number of iterations, the image still consists of very many very tiny cats. The gasket is the limit of this process. But the sequence of cat pictures makes plausible that the limit of this process is the gasket.

As this sequence of images was projected, one at a time, every person in class stared at the screen, many with mouths open. I'd hear gasps, and more than a little profanity. How did this happen? They wanted to know. Then we introduced more general transformations, combined rotation and reflection with translation and scaling. To find rules for more complicated fractals, like the one shown on the next page, is fairly straightforward after enough practice. Still, every year a dozen or so students would tell me that the ability to find the rules impeded their enjoyment of the shapes themselves. Once they learned to see fractal decompositions, the shapes lost some of their beauty.

So we see irreversibility, but would you call this grief? My students certainly didn't. If they described their impressions at all, most said they were sad, and a few said they were annoyed that what had been a mystery was now replaced by the effort to identify reflections, rotations, and translations. These beautiful fractals had become little geometry puzzles. No grief here.

Grief requires more than irreversibility. Grief is irreversibility coupled with the emotional weight of the loss and coupled with transcendence. If the thing lost isn't of immense importance to you, you won't feel grief at its loss. Few, if any, of my students consider geometry to be one of the most important things in their lives.

But I do. Every new bit of geometry I learn, every door that

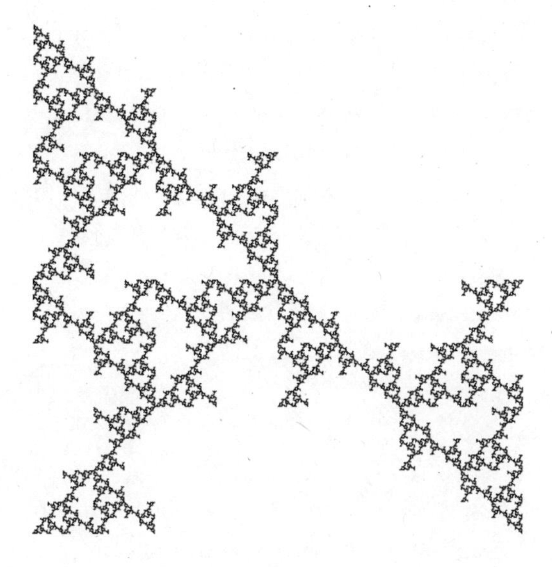

opens in perceptual space, is accompanied by the closing of other doors. I've gradually realized that each of these closed doors forever blocks off a whole world of possibilities; I am unable to look at a new problem in a way unclouded by what I have previously learned. It must be this way: every bit of geometry I learn may enable me to see connections that otherwise would have gone unnoticed, but also it blinds me to connections I might have seen without that bit of geometry. For the loss of these possible worlds, I do indeed grieve. Not as much as I did at the loss of my parents or the cats I've outlived, but still, the loss bites, burns, and is irreversible.

Unless you are as taken with geometry as I am, this argument will likely seem silly. And for you, it will be silly. But my goal here is to help you find analogous parts of your life. The perception of potential loss, of potential grief, may help you better understand ways to approach a situation of actual grief. Here's an example.

For years I ended my fractal geometry course with a story about Henry Hurwitz. In the early 1990s I taught at Union College in Schenectady, New York. There I developed a sophomore/junior–

level math course on fractal geometry and chaotic dynamical systems. Dave Peak, in the Union physics department, had the idea of a course with similar content but less math to introduce non-science students to quantitative thinking. We developed and team-taught this course. When we later left Union, I transported the course to Yale, Dave to Utah State University. Our courses evolved independently of one another, but they have the same roots. The years Dave and I worked together are among the happiest of my life. I wonder what would have happened if we'd stayed at Union, spent twenty more years working together. I grieve that loss. A lot.

Easily the best-known faculty member at Union was Ralph Alpher. A student of George Gamow, Ralph did some of the original calculations that turned the Big Bang model from a suggestive cartoon into solid cosmology with testable predictions. Ralph worked at General Electric's research lab in Schenectady and eventually joined the faculty of Union College.

Henry Hurwitz was a nuclear physicist at GE, where he met Ralph. When Henry retired, he knew that nuclear physics wasn't a hobby he could pursue at home, so he bought an IBM PC and began to look for problems he could solve. The PC software suite included a program to generate images of the Mandelbrot set, pictured on the next page.[22] Could he work on some problem about the Mandelbrot set? Henry asked Ralph, and Ralph sent Henry to talk with me. Several very subtle problems, including one still unsolved despite years of intense labor by brilliant mathematicians, are associated with the Mandelbrot set. But I knew another problem, to derive a proof of a pattern observed numerically by my student Adam Robucci, who I mentioned on page 63. Henry was interested in the problem, and for a couple of months he stopped by my office regularly. I was delighted that Dave Peak joined these discussions. Every week Henry updated us on what he'd done. We'd ask lots of questions, make some sug-

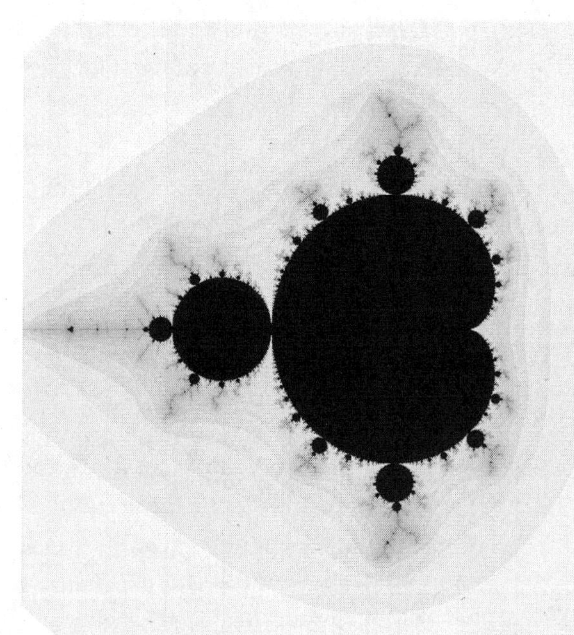

gestions, and Henry would head home. Throughout the week, Dave and I talked about the problem in his office, in my office, and in the hallways between our offices. It was great fun.

Then one day Ralph came to my office and asked how Henry's project was progressing. I said we had made good advances but hadn't found the key idea yet. Ralph asked if we could work a bit faster.

"Why do you think we should work faster?"

"Because Henry was just diagnosed with terminal cancer. He doesn't have many months left."

"That's terrible. What will he do?"

"On the way home from his doctor's office, Henry bought a faster computer."

Well, damn. Now I realized just how much Henry loved to solve problems. Dave and I worked harder. Henry still came to my office every week. He did not mention his illness, so we didn't say anything about it either. And the pieces of the solution began to fall into place.

When Ralph stopped by my office again and asked about the project, I told him we had outlined the main ideas of the proof. A few details remained, but I was sure we'd be able to finish them. Ralph said that Henry wouldn't last past the end of the week. Henry and his wife's kids had come to be with them during Henry's last days. It would mean a lot to Henry to know that Dave and I would finish the project. Would I let him know? Sure, of course.

Ralph left and I thought, Wait, how will I let him know? I'd only seen Henry in my office. I couldn't show up at his home and when his wife answered the door say, "I'm very sorry to hear that Henry will die soon. I've come to talk with him about a math problem." For more or less the same reason, a phone call wouldn't work. At that time, email was uncommon, and in any case it's too impersonal. So I wrote a letter. I told Henry how much Dave and I had enjoyed our work with him, and promised that we'd finish the problem. (We did.[23]) I mailed the letter on Tuesday. The following Monday, Henry's widow phoned. Henry had gotten the letter on Thursday, read it, thought a bit, and announced that he didn't believe Dave and I could finish the solution. He stopped his pain medication and continued to work on the problem, outlining all the steps that remained. Henry's widow said that those last few days they saw Henry at his best, at work on a problem. This was what he loved to do more than anything else. Would I speak at his memorial?

I shared the podium with Ralph and with Ivar Giaever, a Nobel laureate from Rensselaer Polytechnic Institute and one of Henry's coworkers from GE. What I said at the memorial, and what I said at the end of the course, was this: I learned a lot from Henry, but the most important thing was what I learned from observations of Henry at work, namely, that you should do what you love. If you spend your life at a job you hate just to make buckets of money, you're a waste of genetic material. Henry loved to solve problems. I love to teach. The real purpose of education is to sample many fields, to discover what you truly love.

When I finished my career as a teacher, I had no additional advice to give, no notion beyond simple observation of how to figure out what you love. Now I do. Imagine that you were forever barred from work in a field. Would you grieve? Not just be sad, but really grieve. This is a way to tell if you love something.

In the spring of 2016 I gave up my work as a teacher. Poor health kept me from engagement at the level I know my students deserve, so rather than do a half-hearted job, I quit. And this hit me on the head more like a sledgehammer than a baseball bat. I grieved. I still grieve.

But also: the grief that greets me every time I wake from dreams of still being in the classroom shows me that I did spend forty-two years of my life doing what I love. However mediocre my work, it was what I should have done. Geometry and teaching, and cats. I can no longer teach, and every day geometry unwinds itself a bit more from my mind. One door has closed forever, and another has begun to close. Each day my heart breaks for them. But my wife and I still can enjoy a spring morning, an autumn evening. And we still can care for and delight in the company of cats.

The grief of geometry is relevant mostly to geometers, though I hope my story may help you recognize areas in your life that are as important to you as geometry is to me.

On the other hand, in the next chapter I hope to convince you that the geometry of grief is relevant to everyone.

4
Story

The shadows of a broken path

Each of us feels grief in our own way, cut to fit, engraved and inscribed, for our eyes only. Nevertheless, now I'll argue that there is a way for geometry to help us understand our own personal forms of grief. Not a map of stages, but a family of paths, or trajectories, through an abstract space. First we'll see how, then why.

The abstract space is not only the space of emotion. Nor is it only the vastly branched timelines suggested by "The Garden of Forking Paths" by Jorge Luis Borges, where each choice we make selects one branch among all possible future lives.[1] This space, which I call *story space*, is very high-dimensional, maybe infinite-dimensional. It has a dimension for each independent component of the world that can influence your life. This sounds far too general to be of much use. No one, not even the most hypervigilant Force Recon Marine, could keep track of *everything* around them. The usefulness of story space is seen in our changing focus. At any moment we are aware of probably fewer than ten aspects of our lives and surroundings that influence our actions. But as time progresses and situation changes, so do the aspects, the dimensions, that are important. Our path through story space is a fuzzy trajectory confined to a low-dimensional subspace of story space. But which subspace, which dimensions, changes as our lives unfold.

The idea of story space has danced around the periphery of my thoughts for several decades, and was an often-visited side project when poet and journalist Amelia Urry and I discussed examples of fractals in literature for chapter 4 of *Fractal Worlds*.

We'll use the "life as a path" model of story space, though there are other possibilities. In one, time is not an independent variable: our memories and our imaginations take us backward and forward in time, giving some narrative expression to the notion that time is an emergent phenomenon. That we can remember the past but not the future may be explained by the fact that we must clump together data that comes at us; we cannot perceive and process enough detail to remember the future. This is a very tricky idea, well explained in physicist Carlo Rovelli's wonderful books.[2]

Some writers have explored other geometrical representations of stories. An amusing example is Kurt Vonnegut's "Here Is a Lesson in Creative Writing."[3] John McPhee gives an interesting sketch of similarities between narrative structure and geographical feature.[4] Like it or not, we are surrounded by geometry. It influences our perceptions and organizes our thoughts into many categories. And it can help us find patterns we have not noticed.

What are the possible dimensions of story space? For any particular analysis we may focus on only a few, but let's begin with a larger list. Here are some of the dimensions of story space.

- physical position
- emotional state
- physical surroundings
- nearby people

- the current contents of your recent memory
- perceived tasks before you
- the space of actions (story plots)

These are just crude categories: each can be subdivided into more independent coordinates. For example, emotional state can be described by a position along a *fear-comfort* axis. And independently of that by a position on a *calm-angry* axis. And there are many other axes. These are independent because how fearful or comfortable you feel need not have any influence on how calm or angry you feel.

Really? Can we be simultaneously comfortable and angry? From personal experience, I can say with certainty that the answer is yes. I was in seventh grade. Junior high school was about two miles from home, a walk I enjoyed. On this walk one afternoon I saw a (much larger) ninth grader scoop up a cat from the bushes and pull a can of lighter fluid from his pants pocket. Before he even opened the can I hit him with my whole body, fast and hard, behind his knees. Unharmed, the cat ran away and I beat the stuffing out of the kid. I am not proud of the act or of how I felt about it, but the truth is I felt simultaneously angry at the kid and comfortable with my actions to save the cat. A more detailed decomposition of these feelings is impossible, hidden behind the veils of about sixty long years.

Emotional state also involves a position on a *sad-happy* axis, and as many others as the mind can conjure. I've read that we have eight primary emotional states, another source says ten, another twenty. So let's just settle on "lots." If you look at these lists, you'll find they include different emotional states. If this puzzles you, the short answer is that complex emotional states can be decomposed in many ways.

A picture may help to illustrate decomposition, which we can think of as locating a point by different coordinate systems. On

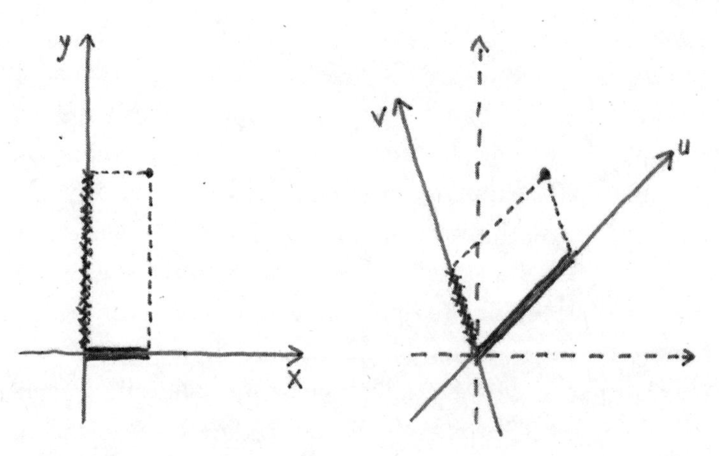

the left we see that the position of every point in the plane is determined by a distance in the x-direction (the solid heavy line on the x-axis) and a distance in the y-direction (the crosshatched heavy line on the y-axis). On the right we see that this same point also can be determined by a distance in the u-direction (solid heavy line on the u-axis) and a distance in the v-direction (crosshatched heavy line on the v-axis). This can be done for any pair of directions u and v, so long as they aren't parallel.

For a physical example, I can describe my (approximate) location in terms of latitude and longitude, or by specifying a street address and city. Both give roughly the same information, but they use really different coordinate systems.

Analogous constructions can be done in higher-dimensional spaces, but the pictures are more difficult to draw. The right way to generalize these notions is the subject of a branch of math called linear algebra.

A useful point, which we'll call the *principle of limited attention*, is that at any given moment we can attend to only a few of the coordinates of our position in story space. Our complete location is always defined, but we are consciously aware only of the projection, or shadow, of this position onto a subspace defined by a small number of coordinates. What's a subspace? It's what you get when you ignore some of the coordinates. For example, the x-y plane is a subspace of three-dimensional space.

The short list of dimensions we gave is just a start. Now we'll be concrete with one limited, if silly, example. On a long walk in the woods, Bill and Steve are fairly comfortable, Steve more so than Bill. Bill hears a noise, possibly a bear, so he becomes more fearful. But soon he sees that the noise is caused by a deer; his fear dissipates and he settles into a comfortable walk, though a bit less comfortable than before he heard the noise. After all, the next noise could be a bear. (See the drawing on the next page.)

Steve is less familiar with woods, so he doesn't notice, or doesn't process, the noise as soon as Bill does. So Steve's fearfulness begins to rise later, and it climbs higher, than Bill's does. Because he also notices the deer later than Bill does, Steve's fear is still increasing when Bill's has begun to recede. Eventually, Steve becomes less fearful, but he remains on edge for a while after seeing the deer.

In a two-dimensional plot with the time axis horizontal and the *fear-comfort* axis vertical, we see that Bill's and Steve's paths intersect at one point. As viewed in this simple representation, at one moment Bill and Steve have the same state of mind.

Now add a dimension, the *calm-angry* axis. Suppose that Steve's position on this axis is at 0 for the duration of this bear scare. That is, Steve's path stays in the plane determined by the *fear-comfort* axis and the *time* axis. The over-crossings of curves and lines should help parse the third dimension depicted in the figure.

Let's also suppose that Bill starts out a bit angry, maybe because Steve has thrown away a lot of their supplies. (If this sounds absurd, consult Bill Bryson's hilarious book *A Walk in the Woods*, the inspiration of this little example.[5]) As time pro-

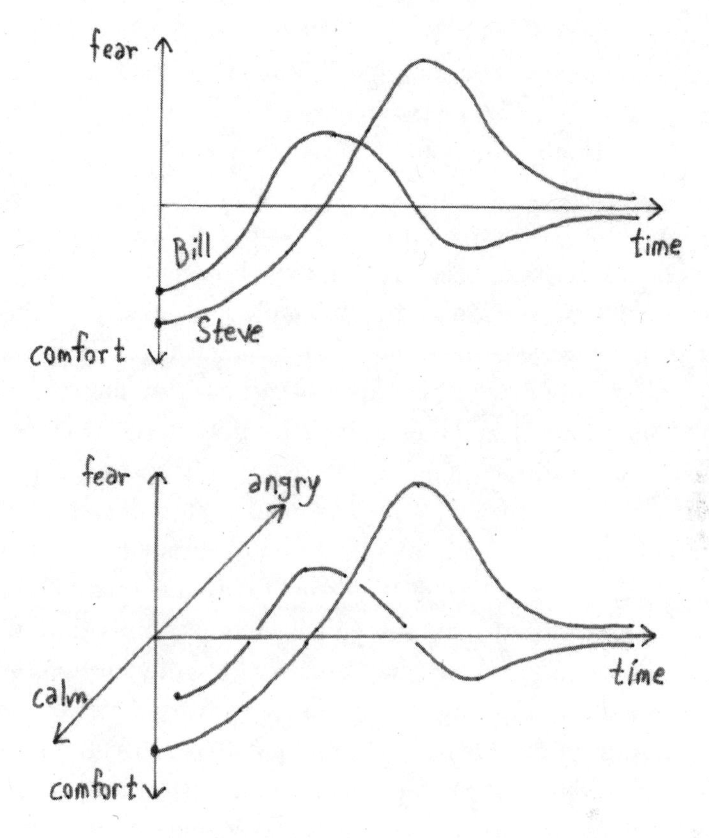

gresses, Bill's anger shrinks, but he remains a bit angry. As we see in the lower sketch above, unfolded in three dimensions, Bill's state of mind never coincides with Steve's.

This simple example shows that by adding dimensions, we may see that paths that appeared to intersect are really disjoint. Conversely, by removing dimensions—that is, by projecting, by looking at a shadow—we can make disjoint paths appear to intersect. But "appear to" isn't quite right: in the subspace on which they are projected, these paths really do intersect.

Why do we care? I'll argue that when viewed in story space, grief is signaled by a discontinuity, a jump, a break, in a path.

Then if we project in just the right way, the shadows of both pieces of the broken path come close together. That is, in this shadow world, grief has been reduced.

Can this work? Can this bit of geometry really help cool the white-hot intensity of irreversible loss?

We'll start with a relation between grief and discontinuity. In high school algebra we learn to distinguish discontinuous paths from continuous. Rather than a mathematical definition, intuition is all we need here. A path is continuous if you can draw the graph of the path without lifting your pencil from the paper. If you have to lift your pencil, the jump from one part of the curve to another is a discontinuity.

Why can we find discontinuous paths in story space? Grief is an expression of irreversible loss. To illustrate the implications for story space geometry, I'll use the example of the death of my mother. The story space of me and my family is partitioned into disjoint subspaces, the *world with Mom* and the *world without Mom*. When Mom died, the paths of everyone in my family jumped from the *world with Mom* subspace to the *world without Mom* subspace. (If you're familiar with subspaces in linear algebra, you can see that I've taken some liberties here. Think of this use of subspace as metaphorical, or just call it a submanifold instead of a subspace.) Two points about this construction need to be mentioned now.

- The jump from the *world with Mom* to the *world without Mom* is irreversible. No life trajectory jumps from the *world without Mom* to the *world with Mom*.
- The story space of *everyone* contains a dizzying array of discontinuities. But only those associated with events of great emotional weight become significant parts of our individual trajectories.

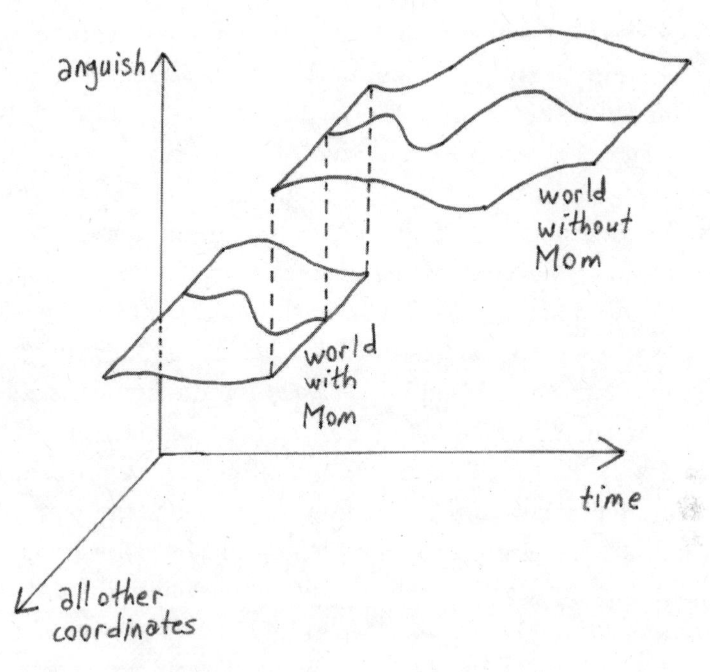

Is grief the only source of discontinuity? Must all irreversibility be signaled by a discontinuity? What about the irreversibility of the first time you watch Peter Sellers in *Being There*?[6] The surprise of the final scene cannot be experienced again. For that matter, every other time you watch Peter Sellers in any film, you'll see Chance the gardener walk out across the water, lower his umbrella up to its handle in the water while the music swells and Jack Warden intones, "Life is a state of mind." Along an appropriate axis—for example, the "Is Chance the gardener god?" axis—this can involve a discontinuity. But is this an axis important for your life, for your thinking? Certainly not as important as the "Is Mom alive?" axis. Discontinuity along an axis with great emotional weight is necessary for grief.

What if we ignore the time axis? Can we find discontinuity

as we change position along an axis that represents something other than time? This is an interesting question. I'll let you think about it for a while. The geometry of story space offers many opportunities for exploration.

We'll end this chapter with a simple example of how familiarity with story space might help reduce the incandescence of grief. Before we get to the example, I must emphasize that I *will not*, *cannot*, provide a general recipe to reduce the sting of grief. I can show you an example, but how, or whether, you can do this depends in exquisite detail on the weights you assign to the axes of your story space.

Rather than build an example that involves the death of one of my parents, I'll construct the example around the death of our first cat, a beat-up little stray who Jean named Scruffy. Our next-door neighbor in Schenectady had cats and fed strays, so cats visited our yard often. Jean liked cats. I did too, but I had pretty bad cat allergies. One little black cat began to spend time in our yard. He would wait for Jean to return home from work at Albany Med, then run to her and rub around her ankles. Jean petted Scruffy, and when she sat on a bench in our backyard, Scruffy would jump into her lap and curl up for a nap. After a few weeks of this, Jean took Scruffy to the local vet for his shots. I stayed home to work. Half an hour later Jean phoned, in tears, to say that Scruffy had tested positive for feline leukemia. Though there is a vaccine, once contracted feline leukemia is incurable, fatal, and terribly contagious. Scruffy would need to be euthanized. Did I want to come to the vet's office to say goodbye. No, not particularly, but you sound like you could use company. I'll be there in a few minutes.

The vet's office was about five blocks from our house. I began walking. Then I thought: Scruffy is a very nice cat, so affectionate and sweet-tempered. Why couldn't we keep him in the basement? I could just stay upstairs. Yeah, that'd work. Then I wondered, How are cats euthanized? The vet gives the cat a shot, I think. What if the vet is preparing the injection right now? So I ran. Summed over my entire life, I doubt I've run a mile. But I ran. Into the vet's office. Where's Jean? Room one. I go into room one. Scruffy is cradled in Jean's arms, and the vet is preparing the shots. (Two are used.) I say, far more loudly than necessary, Stop! Stop! We can keep Scruffy in the basement. What about your allergies? Jean asks. Fuck my allergies!—again, louder than needed. We aren't going to kill a cat because of my allergies. This is a kindness, the vet said, but Scruffy might live as little as six more months. Didn't matter. We'd take care of him for the rest of his life.

So Scruffy moved into our basement, and I stayed out of the basement for six months. Lots of handwashing, plenty of antihistamines. After six months we let Scruffy run around in the house during the day, then put him in the basement overnight. More handwashing, more antihistamines. A year after we took him in, we let Scruffy run around the house all the time. The first night he wasn't in the basement, Scruffy jumped on the bed, got under the covers, and curled up on Jean's shoulder. And he did that every night for almost six years. Love can't stop disease, but can slow it down considerably.

We knew when Scruffy was in his last few days. Feline leukemia had given rise to a true leukemia. Both Jean and I had lost family members to disease, so we were familiar with the general emotional mechanics that accompanies the knowledge that only a short while remains. Experience did not dull the ache of anticipation. Irreversibility struck me even then, now almost twenty years ago. Scruffy would die, he would never un-die, and I could do nothing about it.

We took Scruffy to the vet's office. The vet gave him the first shot, a sedative, then left us alone till we were ready. Scruffy sat up on the table. We petted him and talked to him. He purred, watched us. Then his front feet slid out from under him. We asked the vet to come in. He gave the second shot, and Scruffy was gone. The world turned dark and watery.

Since then we have taken in other strays. And so far we've lost Crumples, Dinky, Chessie, Dusty, Bopper, Leo, and Fuzzy. Each in our vet's office or in the local animal hospital, each to cancer, each broke our hearts. I know of nothing, nothing in the whole wide world, that can relieve or redirect the immediate response to the irreversible change from "Bopper is alive" to "Bopper is dead." For a moment, we are in free fall. The floor won't support our feet and down we go. We are stuck with the first moments of intolerable grief. But starting with Scruffy, I found a way to reduce the residual grief after those initial moments.

But we really don't want to be rid of grief, because the experience of grief is tied in intimate and complicated ways to the experience of empathy. Writing this in the spring of 2020 as I watch mistake after mistake made by leaders of the executive branch of the US government in their wrongheaded efforts to "handle" the COVID-19 pandemic, I can trace many, maybe almost all, of these missteps to two problems: a lack of understanding of science and unwillingness to listen to scientists, and a lack of empathy for regular people whose lives have been turned inside out, or ended, by this virus and its effects. Lack of empathy is one of the main sources of our inability to treat effectively the world's problems. Leslie Jamison's wonderful book *The Empathy Exams* explores many aspects of this.[7] So our goal should be to reduce the pain and misery of grief, but not to eliminate it.

This will involve some visualization in story space. Let's start with a look at projections as shadows. In bright sunlight or lamplight, hold your hand out horizontally and spread your fingers and thumb so that the shadows they cast on the sidewalk or floor are as widely separated as possible. Then rotate your hand, and

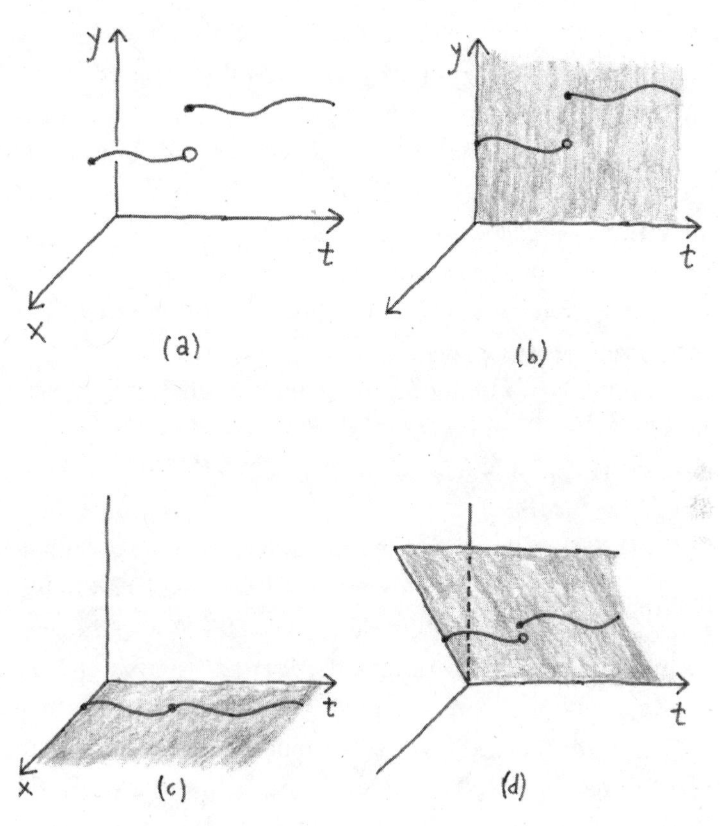

watch the shadows of your fingers and thumb get closer together. Try to reduce the space between their shadows but not overlap them entirely. (For a more amazing example, do a Google image search for "Godel, Escher, Bach." The image shows the shadow of two pieces of carved wood, suspended one above the other and lit from three directions. On one vertical plane the shadows are G above E, on a perpendicular vertical plane the shadows are E above G, and on a horizontal plane the shadow is B.)

Now, on to our model, illustrated in the four-part figure above. In the first sketch, labeled (a), we see a trajectory plotted in three dimensions where the x- and y-axes represent quantities important in story space and t represents time. Note that at the point of discontinuity there is a jump in the y-value of the trajectory while the x-value remains constant.

In (b) we see the projection, or shadow, of the trajectory in the shaded y-t plane, that is, the plane defined by $x = 0$. Because the jump depicted in (a) was in the y-direction alone, the projection in (b) exhibits a jump of the same magnitude as that in (a).

In (c) we have the projection of the trajectory in the x-t plane, that is, the plane defined by $y = 0$. Because x was constant at the jump, this projection shows no jump at all. Consequently, we call this a *trivial projection*. The jump was a change just in the y-value, and this projection ignores that variable. This isn't likely to be useful.

In (d) we see the projection of the trajectory in the $y = x$ plane. Here the see that the magnitude of the jump is smaller. In fact, by adjusting the slope of the plane (the value of m in the plane defined by $y = mx$), we can achieve different length jumps. For the slope m to make sense, we must be able to assign scales to the axes in ways that can be compared coherently. If the size of the jump correlates with the degree of anguish, then this projection can guide our thinking about how we might focus our attention on a small number of factors to ease our suffering in response to an irreversible loss.

Now I'll try to build an example. To reduce the grief at the loss of our cat Scruffy, for a while I tried the approach I use to manage pain. Rather than ignore pain, I focus on it, block out everything else until only the pain remains. And sometimes, but not always, the pain becomes unrecognizable, foreign, and not an issue. I don't know what it is, and the sensation decouples from the usual sense of pain. Would this approach work for grief?

(This approach to pain, which I admit so far has had only limited—but not zero—success, was suggested by a childhood experience. My brother Steve and I shared a bedroom our father built from an extension of the attic of our house. One summer night we had the windows open in hopes of a cool breeze. A dog began to bark. Steve asked what the sound was. I believe he meant, did I know whose dog that was. But I took the question to be exactly as asked and replied, "Just some hound." Steve said, "Hound." I replied "Hooound." We pinged the word back and forth a dozen times. Eventually we both noticed what a strange word "hound" is. We no longer could see any connection between the sounds that make up the word "hound" and the domesticated distant relative of wolves. Repetition and a focus on the sounds temporarily erased the semantic content of the word. I expect this is a common experience. Familiarity overwritten by the utter strangeness of a sequence of sounds signifying a familiar class of object.)

While focus to the exclusion of all else may work for somersaults with words and sounds, it did not dull the pain at the loss of Scruffy. I needed a different approach. Maybe a distraction would work, although I hated the idea. Distraction seemed to deny, or at least marginalize, the importance of Scruffy in our lives. So I thought about the little corners of Scruffy's life: how he jumped from the floor to our shoulders, how he sat at an open window and through the screen made meows that sounded like bird chirps, how he climbed onto a basement shelf and peed on a stack of old *Scientific Americans*, how when I reclined on the couch puzzling over a geometry problem he settled on my chest and tucked his head under my chin. Dozens of other memories shared the stage. Alone, these memories, even his peeing on *Scientific Americans*, magnified the sense of irreversible loss. This wasn't working. But then I thought of other cats, mostly those of our neighbors Bill and Wade. What quirky behaviors I'd observed from them. Rosie was attracted to the smell of cookies and sat patiently in the kitchen, waiting for a corner of a cookie warm from the oven. Princess and her brother Wilhelm chased and retrieved aluminum foil balls, dropped them at our feet and waited till we threw them again. Even though they differed from cat to cat, small traits bridged the gap of Scruffy's absence. He was gone; I could see him only in memories that I knew would fade with age. In this way I'll lose him again and again. But many of the things I loved about him I also can love, and do love, about other cats. The years we spent with Scruffy have refined and redirected my understanding of cats, probably of people too. These ties don't remove the grief that his death has cemented in my heart, but they show me some of how time with him has changed my thinking. One door closes, another opens.

What I think now is that by luck or instinct (but probably luck), I projected the anguish at Scruffy's loss to the space of his small actions and the small actions of other cats. At that time I didn't understand why, or how, this helped, but it did. Now I have a theory.

At least three aspects need to be worked out before this notion of projection can be applied.

- *Non-uniqueness.* Projections to many subspaces, many combinations of emotions, can likely reduce the anguish of grief. Think of the shadow of a pencil held perpendicular to a tabletop. If the light source is almost along the axis of the pencil, the shadow will be short. Moving the light source around a bit will produce short shadows in many directions. Among the many subspaces available for projection, try one that is more familiar.
- *Moving target.* Think of how much your internal model of the external world has changed in the last year. Really, think how much has changed just today. What is important to you now that wasn't important earlier? What used to seem weighty that you now think is no big deal? So far as I know, you can't pre-plan projections. This must be figured out in real time.
- *Calibration.* When you've selected the subspace into which you project your grief, in order to reduce your anguish by a

> given degree, how can you know how much must you attend to the categories that define the subspace? We'll end this chapter with an example that illustrates an approach to this question. In story space this is a question of the slope of the plane (or its higher-dimensional analogue) to which you project.

Now I don't know if this approach will be of use to someone whose thinking is not primarily geometrical or visual. But thoughts of projections onto other planes have helped me sort out this approach to reduce the pain of grief. The first two points you must work out for yourself. I encourage you to visualize how the various factors relate to one another. Let the shapes twist and turn in your mind. Initially, this may be demanding, but I do believe that the clarity, and freedom, of geometrical thinking are without equal.

Benoit Mandelbrot, with whom I worked for twenty years and who catalyzed my move from Union College to Yale, had an amazing ability to think geometrically. He told a story about how when he took the university entrance exam in France one of the problems was a particularly difficult triple integral. After the exams were marked, and Benoit had won admission to any school in France that he wanted, his high school math teacher told him that in the entire country only one student had solved the big math problem, and that student had been in his class. He (the teacher) couldn't evaluate that triple integral in the time allotted. Had Benoit solved it? Yes. But how? Benoit said, "I moved the shape around in my head and saw that with an appropriate change of coordinates, the triple integral reduced to finding the volume of a sphere, and I know the volume of a sphere."

When Benoit first told me this story, I felt even more intimidated, though Benoit and his wife Aliette were never anything other than kind to Jean and me. But eventually I recognized that

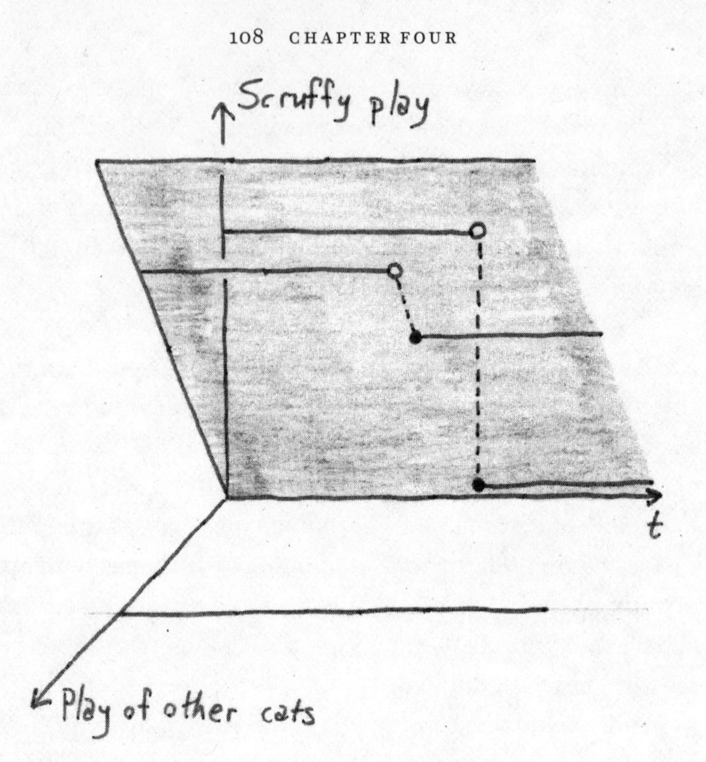

I didn't need to be able to do geometry at Benoit's level in order to do some geometry. We each have our own skills. Cultivate yours.

In the next chapter we'll explore an approach to my third point, calibration, based on scaling. For now, though, I'll give a simple example of direct calibration. In the sketch of this calibration, the vertical axis, labeled *Scruffy play*, represents my pleasure at watching Scruffy's inventive play. The axis labeled *Play of other cats* represents my pleasure at watching the inventive play of other cats. In the subspace of *Scruffy play* and time t, my path has a large discontinuity at the moment Scruffy died. After Scruffy died the pleasure didn't drop quite to 0 because memories can give some joy.

In the subspace of *Play of other cats* and t, my enjoyment remains constant, not as strong as that of watching Scruffy play, but more, certainly, than my memory of Scruffy's play.

The shaded plane is a subspace where I think sometimes about the play of other cats and sometimes about Scruffy's play. Consequently, the drop after Scruffy dies is not as large as it was in *Scruffy play-t* subspace.[8]

Careful calibration would be difficult. I'd need a way to compare how much I enjoyed watching each cat play. For any real application, something along these lines must be done. A very simple surrogate is the count of the relative number of times I remember Scruffy's play and the number of times I watch other cats play. I must emphasize that watching other cats play is not a distraction from memories of Scruffy but a reminder of his life, that some versions of his behaviors are expressed by other cats. But I don't want to forget Scruffy, so I attend to the ratio of the time I remember Scruffy to the time I watch other cats playing, and make sure the ratio doesn't get too low. With this approach in mind, the calibration may self-correct.

Any application of these ideas to the more complex relations we have with people we have lost will require a much more sophisticated approach. This is just a simple illustration of how calibration can work.

I've used geometry because it is familiar, has worn tracks in my mind for over six decades. Consequently, the dance of shapes is the basis of my approach to guide you to useful combinations of perceptions. Other paths may take you to the same goal. Perhaps your imagery of the world is more auditory or tactile. Is your day marked by bits of songs? Then music may guide you in how to find what corresponds to these projections.

Or stories, or movies, or chess, or cooking, or dance, or spending lots and lots of time with cats. Whatever is important to you should be able to guide you to projections that gentle your grief. But I think this can work only if you have real passion for the path you explore. You may find ways, undreamed-of by me, to discover combinations of perceptions that can reduce the violence of grief.

5
Fractal

A day is a laboratory for a life.

We've examined the self-similarity of some shapes, specifically the Sierpinski gasket. The right isosceles version of the gasket, seen in the top figure on the next page, is made of three pieces—lower left, lower right, and upper left—each a copy of the whole gasket scaled by a factor of one-half. This sets in motion a process that can continue forever: each of the three pieces is made of three still smaller pieces, each of which is in turn made of three even smaller pieces, and so on. Shapes that exhibit this kind of symmetry, symmetry under magnification, have been familiar to, and constructed by, artists for at least a millennium.

More natural fractals can be made by a process called *decalcomania* (see the lower left image on the next page). Paint is flattened between two surfaces. When the surfaces are pulled apart, air intrudes and produces intricate branching patterns. This technique has been known for at least several centuries but saw its greatest proliferation early in the twentieth century with the work of Max Ernst, Óscar Domínguez, Boris Margo, and Hans Bellmer, among others. The complex branching contributed a surreal, dreamlike quality to their paintings.

Natural fractals abound in the physical world. Clouds, mountain ranges, coastlines, river drainage networks, all have no natural scale. Without other clues you can't tell if you are looking

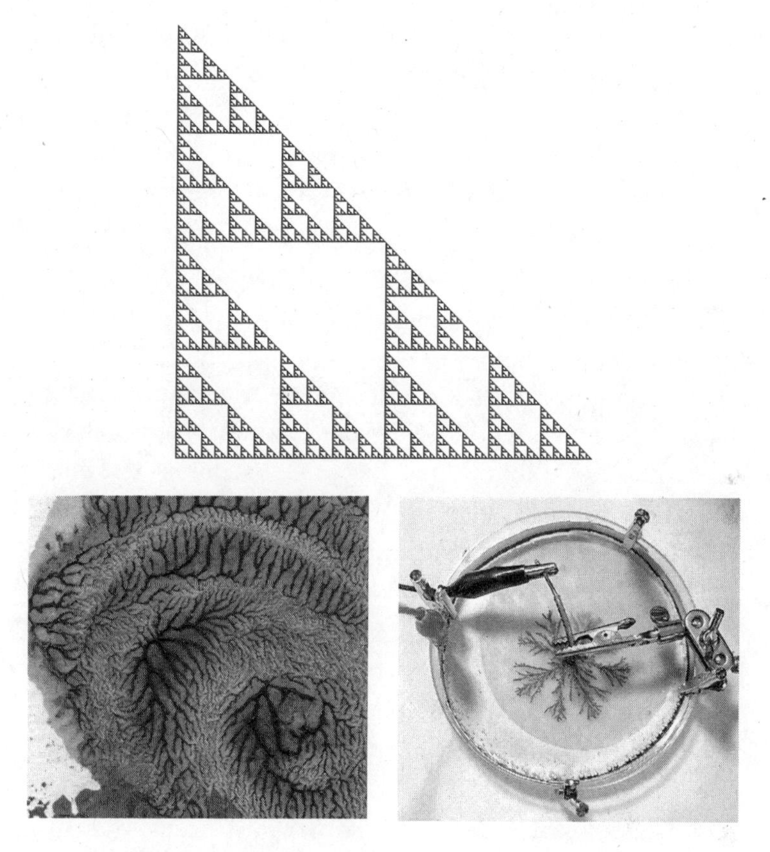

at, for example, a small cloud from nearby or a large cloud from far away. Similar structures occur on many levels. The lower left photo above offers no clues to scale. In the second, the alligator clips betray the size of the apparatus: we've grown a fractal dendrite by passing a small electrical current through a zinc sulfate solution.

Fractals occur is literature, too. For example, José Saramago writes this about the geometry of the cemetery in his novel *All the Names*:

> There was a moment in my life when, without my actually noting the phenomenon, I found myself deeply involved in something

as mysterious as fractal geometry, of which, with apologies for my ignorance, I had absolutely no prior knowledge.[1]

Saramago describes the arrangement of gravestones as a branched tree, the oldest graves in the trunk, the newest graves at the branch tips. The fractal geometry of this cemetery was pointed out to Saramago in 1999 by the Spanish mathematician Juan Manuel Garcia-Ruiz. Indeed, fractal geometry is an immense field.[2]

We've described these examples to give a wider picture of fractality, so you won't look at the (necessarily messy) physical fractals in our lives and say, "Wait a minute, that doesn't look like a gasket." We're looking for patterns that repeat, approximately, across scales in space or in time or in some more abstract setting.

Here's an example of similar structures across time scales.

- Consider a day. You wake up in the dark, think of what you'll do that day. By morning you're well into the day's work; by evening you've finished and reviewed what you've done with the day. Then day's end and sleep.
- Consider a year. In the dark of winter you think of what you'll do that year. By spring you're well into the year's projects; by autumn you've finished much of the work and reviewed what you've done with the year. Then winter, year's end, and rest.
- Consider a life. As a child and adolescent you develop the tools for the work you'll do. By adulthood you're well into your life's work; by old age you've retired and reviewed what you've done with your life. Then life's end and the sleep of nonexistence.

Certainly this crude sketch cannot capture the rich and varied detail of a life, but it does give a glimpse of patterns that repeat across different time scales. Beyond our natural affinity to find patterns, why should we care? Because a day provides a laboratory for a year, and for a life. To try to influence trajectories on longer time scales, experiment with analogous changes for a day. How do you find analogous changes? The beauty of short–time scale experiments is that you can try many changes and note their short-term effects. Fractality offers a stage to test hypotheses on small scales that can influence circumstance on large scales.

I'll argue that grief occurs on different scales, both in time and in anguish. If we understand how to deal with grief on small scales, can that help us deal with grief on large scales?

In fact, we've already discussed a laboratory for experiments on grief: our study of geometry (or replace geometry with the strongest focus of your curiosity) in chapter 3. Pick an area of geometry new to you. In chapter 3 I used fractal geometry because the details of that subject are unfamiliar to many people. It is the most visual geometry that has a chance to offer surprises to almost everyone. (In fact, maybe everyone. When someone showed Benoit a new calculation or experiment or observation about fractals, his delight was palpable. This brilliant, usually reserved, man at the height of his career was transformed into a little kid who'd looked up into the night sky and seen a meteor, a bright streak of light where a moment ago had been darkness.) We'll continue with fractal geometry and discuss a consequence of this study that does surprise many people: that dimensions needn't be whole numbers.

For examples take a line segment, a gasket, and a filled-in square (shown at the top of the next page). If we double the height and width of each shape, we obtain two, three, and four copies of the original shape. The line segment is one-dimensional and the square is two- dimensional, and when we double the height and width of these shapes we obtain $2 = 2^1$ copies of the line seg-

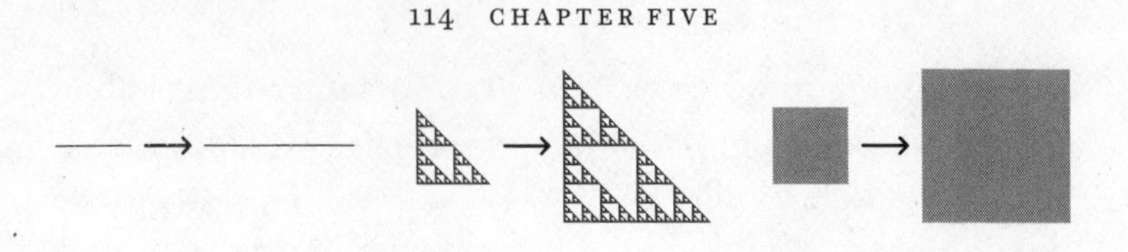

ment and $4 = 2^2$ copies of the filled-in square. For these examples, and it turns out for all self-similar shapes, the dimension is an exponent. So the dimension d of the gasket is determined by $3 = 2^d$. Now $2 = 2^1$ and $4 = 2^2$, so the dimension of the gasket is more than 1 and less than 2. As its size increases, the gasket grows more rapidly than a one-dimensional line, but more slowly than a two-dimensional filled-in square.[3]

The gasket occupies a world between the one-dimensional and the two-dimensional. Early in their efforts to understand this, some of my students thought a narrow strip in the plane would be between one- and two-dimensional. The strip doesn't occupy the whole plane, so, they imagined, it is less than two-dimensional. And it's fatter than a line, so the strip is more than one-dimensional. The second statement is close to correct, because of what's called the "monotonicity" of dimension: the dimension of the part cannot exceed the dimension of the whole. The first statement is more problematic, because a shape—every shape—that has an area is two-dimensional. A narrow strip in the plane is two-dimensional. But the gasket has infinitely many holes and their areas add up to the area of the big triangle, so the area of the gasket is zero.[4]

Fractal dimension has many applications, including as the first repeatable measurement of the roughness of physical objects. Extending the notion of dimension to psychological or perceptual spaces is tricky, but here's a final thought, more a guess or maybe a wish. The self-similarity of grief suggests that we can use small losses to test how to adjust to larger losses. Can we measure the dimension of a projection and read that as a guide, however crude, to the strength of the connection between the larger and smaller losses? Right now, I don't know. But maybe, eventually.

Here's a preliminary exercise: If you lived in a world where the dimension of space were not a whole number, how would your surroundings appear?[5] What if the dimension of time were not a whole number?

More questions than answers, but these aren't even really questions, just dreams.

When you first see this idea and understand its implications, your view of the world rotates. Watching my students get the point, waves of amazement for most, waves of vertigo for a few, wash across the room.

This is why teaching was such a remarkable experience, why the only reason I'd miss a class was being in the hospital. Even now, half a decade after I stopped teaching, I still dream about it. I wake up and think I made a dreadful, dreadful mistake when I retired.

The complexity of visual images, the roughness of tree bark, the fluffiness of clouds, the density of tree branches or fern fronds, all of these now give a sense of number. When you first see this, you think, "For my whole life up until now, I've not seen this way to understand the complexity of the world." And now you've learned a new way to measure it. But this sense of surprise fades over time, the shock of the initial revelation does not repeat, and you can grieve for the irreversible loss of the awe of opening your eyes for the first time in this way.

Can we recover echoes of this feeling? Maybe. We can project the surprise at finding dimensions that are not whole numbers into many different settings. Generalize the simple formula where all of the pieces of the fractal have the same size, to self-similar shapes with different scaling factors, to fractals where only some combinations of transformations are allowed (we saw an example in chapter 1, on pages 28–29), to fractals in which the scaling factors are selected randomly, to fractals where the scalings are nonlinear, and on and on. The simple formula for fractal dimension (introduced in the appendix) can be generalized to more and more settings, and all these versions carry the thumbprint of the original formula. Each one of these extensions is a small surprise, gives a twitch similar to the original shock of non-integer dimensions.

And as we assemble this collection of similar formulas, we see that they all are shadows of a larger picture. By projecting the grief at losing the initial shock of non-integer dimensions into different spaces, we can find some alleviation of that grief by discovering small echoes of the original surprise. But look at what's happened in this example: all of these projections have shown us a deeper underlying pattern. Can this reversal be transported to other encounters with grief, grief at the loss of a person or an animal?

As far as I can tell, all forms of grief are similar: they are characterized by the irreversible loss of something or someone of great emotional weight, coupled with a hint of transcendence. The magnitudes vary greatly, of course. The grief at losing the cognitive realignment caused by introduction to non-integer dimensions is less than the grief at losing a pet, which is less than the grief at losing a parent. Griefs differ in degree, but not in kind. Or so it seems to me.

And this leads to another point: each grief has many subsets, subgriefs really. When we lose a person, we also lose the possibility of new instances of things the person did. Each act consists of many pieces, subacts, and we lose the possibility of new instances of these. And so on. If all subgriefs are similar, then grief is self-similar. Recognition of this self-similarity may help us identify useful projections to soften grief. This is too abstract. Let's illustrate the main idea with an example that's more visceral than computing dimensions for a collection of increasingly irregular fractals.

I'll return to a more universal grief, the loss of a parent. This time my father. Mom's death was unexpected: a stroke and she was gone. Dad lived for seven years after Mom died. He had health issues: diabetes, heart bypass surgery, emphysema, and asbestosis. The last two came from his work in the Newport News shipyards early in World War II, where one of his first jobs was to spray asbestos between the walls of the shipboard munitions stores. Eventually he worked as an electrician's assistant and installed many of the landing lights on the aircraft carrier *Yorktown*. But his work with asbestos insulation, where at the end of a shift he looked like a snowman, a snowman without a respirator, eventually did him in. Well, fifty years of cigarettes didn't help.

Dad stayed on in the family home, much of which he'd built and upgraded himself, for five years after Mom died. He learned to cook a bit, to do laundry, to clean some. Dependence on an external oxygen source kept him closer to home. He asked my sister to find an exercise program for him, and she enrolled him in a nearby Silver Sneakers class. He gave up driving, except around town. St. Albans is a small place, so he did not travel far. But he was losing ground. He got confused, didn't always recognize people he knew, including me. Then he became nervous about living alone, began to sleep with a loaded pistol under his pillow. One night he was awakened by a sound on the deck behind the house. He got up, took the gun from under his pillow, and walked through the house to the back door. He turned on the deck light, opened the wooden door and there, separated from him only by the glass storm door, he saw a naked man holding a gun. Dad raised his gun, the man on the deck raised his gun. Then Dad realized he was about to shoot his reflection. (That he slept naked was not something I'd known till he told the story of his showdown with his reflection.) Soon after that he asked to move into assisted living. My sister found a good place for him. Dad sold the house and moved to the place Linda had found.

There he stayed for eighteen months. During visits other family members ran errands for him or brought over old friends. I've never learned to drive, so I just stayed with Dad, him in his rocking chair, me on the couch beside the chair. Usually he'd put a cowboy movie in the DVD player, though occasionally I'd sneak in an old Jimmie Stewart or Alfred Hitchcock movie. We'd talk a while. Soon Dad would drift into a nap. As far as I could tell, he watched only a small number of movies, over and over. If anyone asked about this, he replied that he saw something new every time he watched. Probably he was awake for different parts each time. I'd get him to tell stories about his childhood in Rosedale, West Virginia, during the Great Depression, or about his time in the Navy in the Pacific during World War II, or about his early postwar life and how he courted Mom. He had about two dozen favorite stories, so on most visits I heard replays of his greatest hits. Occasionally I'd hear something new, but usually not.

Then early in 2016 Dad took a sharp turn for the worse. Moved to hospital for a few weeks, then to hospice, then he was gone. The last few days he did not seem to see the same reality as that of the people around him. He told Linda that one night he'd talked with his wife (dead seven years by then), told her how well their kids were taking care of him. He said that Mom replied, "Well, what did you expect?" A few more days there, and then one morning he was gone.

Because Dad was a Navy veteran, he was entitled to a military funeral. I'd been to military funerals before, so knew what to expect. Because Linda had done so much to take care of Dad, ever since Mom died, but especially these last few years, Steve and I asked the Navy chaplain to have the flag presented to her. I was okay when the chaplain read the Sailor's Prayer, when the sailors so precisely and solemnly folded the flag that had draped Dad's coffin. And I was okay when one of the sailors knelt to present the flag to Linda. She did not expect this, and this was when she broke down. Over the last few days I thought I'd shed every tear I had, but when the chaplain said, "You might want to cover your ears. It's about to get loud," I found that I was mistaken. Seven retired sailors and soldiers, three shots each, then the bugler played taps, the saddest music ever written. My breath abandoned me, a Niagara of hot water from each eye, and I seemed to squeak rather than wail.

After I'd recovered a bit, I thanked the chaplain. She told me that sometimes she had trouble finding seven people for the rifle salute, but because Dad was a World War II veteran, she'd had plenty of volunteers. "Unless you've been in the military, you can't understand what an honor it is to show respect to someone who saw active duty in World War II. Your father was a hero." I knew he was a hero to me and to Linda and Steve, but I'd never thought others might feel this way. And somehow still more water leaked out of my eyes. How is this possible? Where is all this water when I'm not weeping?

On the return to Connecticut, Jean and I talked a lot about things we had done with Dad. About how much he'd worked on our house; about how shortly after Jean and I were married, Dad and Mom had taken us on a tour of West Virginia, to "show off" the state to Jean, who grew up in the Upper Peninsula of Michigan; about evenings on his deck after Mom had died, watching the stars appear above, the fireflies below, how one evening the conversation drifted out to dreams, regrets, the slippery nature of past and future. These memories help. For a moment I am working with Dad to replace the breaker box in our house; or riding with him, Mom, and Jean past the radio telescopes in Greenbank; or enjoying the first chill of a summer evening and wondering how my old dad suddenly saw, and could talk about, ideas three levels deeper than I'd ever heard him discuss. Had I misunderstood him these last sixty-five years?

These memories helped for a moment or two, but were gone

far too quickly. And yet, soon after the heartbreak of Dad's funeral, the pain of grief receded a bit. Why? I certainly didn't love Dad any less than I'd loved Mom, and Mom's loss burned for years. The difference, I believe, is that with Dad I'd had some practice. He didn't vanish in a flash as Mom had done. Bits of Dad disappeared in the years that led up to his death.

First, his respiratory ailments eventually forced Dad to stop working in his shop. At some point he knew he'd not go in there again. I knew it, too, and suffered from the irreversibility of the situation. Dad and I had worked on hundreds of projects in that shop. Until I went to college, a corner of Dad's workshop was my lab. When I knew he'd not work there again, I thought of small things we'd done. Repaired a neighbor's lawnmower, built picture frames and cabinets, built jigsaw puzzles and wooden toy cars for neighborhood kids. I focused on the actions, not the feelings, and imagined other people helping their neighbors in similar ways. I saw what Dad did, occasionally with my help, as part of a larger picture. Even though he would not do this again, the idea, the movement, of neighbor helping neighbor, to which Dad belonged, would continue. Projection to the space of *neighbor helping neighbor* eased the grief at the closing of Dad's workshop.

A similar, but stronger, pain occurred when Dad's house sold. Linda, Steve, and I had grown up in that house. So many wonderful memories. In the evenings, three little kids piled up like puppies around Mom; Mom read stories while Dad peeled and sliced apples and passed them around. Lots of laughter, some arguments, some tears. Many stories, many meals, many conversations. And the house grew as we grew. Dad added one room, then another. Mom sewed curtains, planted a garden. The shape of the rooms, the geometry of the spaces, was folded into our lives. Then Mom died. Then Dad moved to assisted living and sold the house. This, too, was irreversible. We never would live there again. And so that loss, too, was a source of grief. The couple who bought the house were expecting their first child. They made an offer during their first walk-through. Dad's upgrades were thoughtful and solid; he was pleased, and flattered, to get a good offer so quickly. When I spoke with Dad right after the sale, he said the house needed a family, that he was happy another little kid would grow up there. He was right. To soften the grief of the loss of our home, project the small details about our life in the house onto how another family will live there.

When Dad died, after the initial heartbreak of the funeral, after the twenty-one-gun salute wrung from my eyes tears hotter than I've ever imagined, I thought of how we got through the grief of the loss of the workshop, and of the house. Project to the space of small details and of interactions with other people. All the work Dad did to help other people—home repairs for neighbors, home construction for family and friends, always happy to listen to stories, to share his own—had an impact beyond the people he helped. His actions, and Mom's cooking and sewing for others, were examples of kindness, of generosity. In small ways this would spread. That was Dad's true legacy, Mom's true legacy. They are gone. I'll never see them again, never talk with them again. But their slow, gentle, reliable work has helped people find that path. "Small steps," Dad told me when I was stuck on a problem. By small steps they left the world better than they found it. And really, this is the best almost any of us can do.

These projections likely will not be effective for everyone. So far, I have not found a projection that is universally applicable. Maybe there is none, or maybe I'm not bright enough to find it. As I understand this approach now, implementation is personal. The heavy lifting is yours to do. The geometry of projection has been a visualization tool useful to me. But if you understand the concept and don't care for geometry, you needn't visualize anything. If you can find smaller scale samples of grief, use those as a laboratory to investigate effective projections. Then use the self-similarity of grief to scale upward. If the smaller-scale samples are components of a larger grief, as they are in my examples, then the reason I call this self-similarity of grief should be clear. I hope you can adapt this approach to your circumstance.

Lara Santoro, a journalist and author who spent years reporting on the HIV/AIDS epidemic in Africa, and consequently a person who has seen more grief than most of us, has found a way to keep from being overwhelmed. She calls this "zooming out." Instead of being caught in the moment, tied completely to what she sees, hears, and feels, Lara steps back mentally, watches herself in the middle of all this heartbreak. That extra level of awareness applies a filter to empathy that could crush. Yes, she still is immersed in images of the despair she would feel were she in those situations, but observing her own situation from outside means that despair no longer is the whole story. And this is enough. It still hurts—terribly—but now it is tolerable. Lara traces this idea to Spinoza, who said that suffering ends, or at least is dulled, when we form a clear mental picture of it.

The method of projection I'm suggesting is mostly concerned with seeking levels within grief, substructures of grief. Lara looks out rather than in. Her approach bears a superficial resemblance to my approach of noticing a deeper underlying pattern in the various forms of dimension. But whereas I stay squarely in the world of ideas, Lara imposes a personal element, maps an awareness of herself into the process. Hers is a brilliant approach.

This brings us back to stories. I don't want to end on a pessimistic note, but because one more chapter remains, here I can give a warning, a plea for you to be better than I have been, to avoid the irreversible heartbreak of looking back over what you've done with your life and finding mostly the bitter ashes of regret. When I began this project, I wasn't sure I'd include this. Now, it seems unavoidable. If I change my mind, you'll never know.

First, though, here is Helen Macdonald's representation of how a life might unfold differently than expected:

> There is a time in life when you expect the world to be always full of new things. And then comes a day when you realize that is not how it will be at all. You see that life will become a thing made of holes. Absences. Losses. Things that were there and are no longer. And you realise, too, that you have to grow around and between the gaps, though you can put your hand out to where things were and feel that tense, shining dullness of the space where the memories are.[6]

At every stage in my life, I have taken the comfortable choice. Mom wanted me to become a medical researcher, but I didn't take that path, which would have been very challenging and worrisome. I knew I'm not bright enough to do great work, but I could have done something useful. Rather, I studied physics and math, tried to convince myself that abstract work was in some way "better" than applied work. That is nonsense; it's a way to hide from responsibility. Had I worked in biomedical research, mistakes I might have made could have hurt people, maybe very badly. But if I made a mistake in the proof of Green's theorem in a third-semester calculus class, no one died.

I was smart enough to learn some math, but not to do any significant research. So instead I focused on teaching. And because I'm confused much of the time, I was sensitive to confusion in my students. Usually I could see it, make adjustments before their confusion crossed the "time to ask a question" threshold. And I convinced myself that I taught in order to help my students. Maybe I did. To be sure, teaching is a noble calling. My sister taught second and third grade in rural Ohio. She is a hero. I am a lazy bastard.

All my energy went into refining my craft as a teacher. Then in my late fifties, cognitive troubles began to appear. The dance of ideas in the classroom no longer was so clear. In classes I'd taught for over a decade, rather than five minutes to go over the notes for the day—reviewing which examples, what theorems, which applications, I'd be presenting—I found myself going over the notes again and again in the hour before class, hoping enough would stick. Sometimes it did, sometimes it didn't. Teaching is the only thing at which I wasn't terrible, and I had to watch this one skill slowly dissolve. Neuropsychological evaluations and PET scans revealed real problems. I wasn't just getting old and tired; I was becoming transparent, a prelude to becoming absent.

This decline was accompanied by some ironies. I'll mention only two. No need to bore you with more of my troubles. I was invited to give the "Thinking outside the box" lecture, on fractal geometry and the complexity of life, in the 2014 Vienna Biocenter PhD Symposium, an event organized by graduate students. A few years earlier, when I still was reasonably confident that I could do a decent job, I would have loved to go. I've not visited Vienna, or any part of Europe. This would have been a good opportunity. But by 2014 my mind already had begun to disappear. With great sadness, I declined and recommended another speaker who worked in these fields.

And the final possible irony: when I returned to my office after giving the last lecture of my career as a teacher, I found an email from the Mathematical Association of America, asking me to give one of the invited addresses at the 2017 MathFest in Chicago. I thought about it overnight. Jean and I like Chicago and know it reasonably well, though we haven't been there for some years now. But by May of 2016 I did not feel confident that I'd be able to do a competent job a year later, so again with great sadness I declined.

Did I grieve for the loss of skills cultivated for decades? For the need to turn down invitations to what would have been amazing opportunities to share some of what I'd learned from working with Benoit? Yes, oh yes.

In her novel *Strange as This Weather Has Been*, Ann Pancake gives an electrifying description of grief at the loss of what life could have been:

> I learned what it is to grieve your life lost while you're still living, and I learned that there are few losses harsher than that. It was grief beyond anything I'd imagined. I can still feel sometimes that dry raw socket. The slash, then the body-burning pain.[7]

The self-similarity of my life choices—in matters small, medium, and large I've always taken the safe, comfortable choice—has generated a self-similarity of grief: I regret the small, medium, and large choices I've made. Small-scale grief: why did I take another astronomy course instead of a genetics course? This should have given a preview of large-scale grief: I could have helped find cures for diseases, or treated patients. Instead, I covered chalkboards with diagrams and equations, and tried to explain bits of how geometry unfolds nature. But I hadn't seen, let alone recognized the implications of, this scaling back then.

Has stepping back and viewing myself in the midst of heartbreak helped dull the sting? Yes, a bit. Mostly, this point of view has brought some relief to the despair over paths not taken. No matter what I'd done, eventually I'd have wound up here, and there are careers far more heartbreaking to give up than four decades as a teacher.

Will you make better choices with your life than I have made with mine? I'll not know, but you will know.

6
Beyond

The glowing audacity of goodness

JUDEA PEARL

We've seen that an appreciation of geometry can guide a redirection of our perceptions to blunt the impossibly sharp point of grief. We'll end by moving in another direction. Rather than project into different subspaces of our perceptions, we'll project into the space of our actions. And we'll apply scaling here, too.

First we must address a point about individual differences. When my mom and dad died, first there was paralyzing pain. When that abated, I was smacked up the side of my head when I saw something that would have interested Mom or Dad and my thoughts of how to tell them were interrupted by "Wait, they're dead." When those feelings were no longer so common, I began to fill in the very large gaps their absence left in my world. But then, I began to dream about them. Even now, I still dream of them, usually something mundane: a short trip, work with Mom in her kitchen, work with Dad in his shop. When I wake up, I realize afresh that they are gone, and I have a short, intense burst of grief, usually accompanied by a string of profanity.

My sister Linda also dreams of Mom and Dad, but when she wakes, she's happy because for her the dreams have been another visit with them. Linda is about two and a half years younger than I am. We have many shared experiences; we both became teachers. But we react to dreams of our dead parents in quite different ways. The point: you can't use your feelings to guide how you think another person feels. Observe without prejudice.

When you lose someone, people around you will say well-meaning things that may upset or annoy you. Platitudes are all most people can offer in these situations. A small projection into the space of actions: even if what someone says bothers you, be gracious. To say what you really feel would hurt them, so act in a way that may make them feel better. You're the one who's grieving, but turn your attention outward to help the people trying to comfort you. If they've brought food, compliment them for their thoughtfulness. Partly this will distract you, but mostly you'll feel better because helping people makes us feel better.

(When I talk with a person who has just lost someone close, usually I ask if I can do anything to help. Observe and guess at what needs to be done. Is there an errand I can run? A call I can make? Specific offers are better than generalities. And then ask if they would share a story about the person they've lost. These small gestures may not help. An offer to wash the dishes may be met with, "How can you think of dishes at a time like this?" Asking for a story may be met with sobs. Use all you know about the grieving person to inform your guess as to what they need, and be prepared to absorb the consequences if you've guessed wrong.)

Grief gives us the opportunity to project outward into actions that can help others. These examples are small steps, but even they reveal some of the reasons that group selection amplifies the actions we can take to recalibrate our lives after a loss. Can these modest actions scale up? Are there large steps?

In fact, sometimes grief gives us the opportunity for acts of breathtaking good. My grandparents and parents have died, and those losses were terrible. But for a parent to lose a child is so much worse. I have no children, so never could experience this directly. I think I cannot imagine the incandescent heartbreak of losing a child to disease or to accident, much less to murder, to a political assassination. Just the contemplation of that grief is paralyzing, even though for me this contemplation is abstract: never having given a parent's love, I cannot know a parent's loss. But I can appreciate, however indirectly, the magnitude of such loss.

Judea Pearl is a brilliant computer scientist who developed a causality calculus, described in his books *Causality* and *The Book of Why*.[1] Among other thing, causality calculus resolves a puzzling set of calculations called Simpson's paradox in statistics. Judea Pearl's son was the journalist Daniel Pearl, kidnapped and murdered in Afghanistan in 2002. Of the responses to this worst of all possible nightmares parents can have, Judea, his wife Ruth, and their family and friends formed the Daniel Pearl Foundation, whose purpose is to promote cross-cultural understanding. In the face of so much evil, this is the most heroic response I can imagine.

In a January 2009 journal entry, the late film critic Roger Ebert wrote that he never cries at sad moments in the movies, but only at scenes of goodness, moments that give a feeling of uplift, what he called elevation.[2] "I am moved by generosity, empathy, courage, and by the human capacity to hope." The response of Judea and Ruth Pearl expressed vast amounts of these qualities, and yes, when I read about their actions, I felt a stinging in my eyes, tightness in my throat, rapid breaths. The next day, after I fed the feral cats who live in our yard, the force of the Pearls' response hit me. I sat on the porch steps and sobbed. Even now, I experience some emotional challenge in writing about the choice they made: when confronted by absolute horror, to pivot to a celebration of "the glowing audacity of goodness" in their son's life.[3]

I can't begin to describe the path that led the Pearls to this action.

Rather, I'll give an interpretation that uses the techniques we've developed. Danny was interested in music, so project the loss into a space that includes his love of music. The music will persist, so Danny's influence, his awareness, will leave echoes. His loss, always before his family's eyes, will recall details of his life. They will have no new experiences with him, but memories of him can be viewed from many perspectives, understood in evolving ways. Projecting the memory of Danny into the space of his actions and interests casts these memories in new lights. But take a step back. What did Danny want to do? Can we help people who didn't know him experience his intentions? The answer, beautifully and breathlessly, is yes.

Death closes the door to further experiences with those we have irreversibly lost. But grief opens a door, maybe just a crack, to let us remix memories, see actions in a new way. Let us think what the person who has died would want us to do. Examples are familiar: "In lieu of flowers, the family suggests a donation to . . ." This is good, this is wonderful. A cause dear to the person who has died gets a boost in their memory. Their influence still is felt.

And for a few people, grief flings open wide the temple doors, gives them a way to step back and do startling good.

Does grief have an evolutionary basis? Move up a level to the evolution of society. Grief can catalyze acts that can help multitudes.

Maybe this is the best answer to our pain: grief can give us agency to take a bold step.

Appendix: More math

A look under the hood

Here we'll fill in a few details that weren't needed earlier if you were willing to take my word on certain points. You'll need to be familiar with a bit of math, high school algebra in most places, to follow these arguments.

HOW MANY CUBES BOUND A HYPERCUBE?

The unit square S in the xy-plane consists of all points (x, y) with $0 \le x \le 1$ and $0 \le y \le 1$. To find the pieces of the boundary of S, set one of the coordinates to an extreme value, 0 or 1, and let the other coordinate vary over the entire range $[0,1]$. So the boundary of the square is four edges, each a line segment.

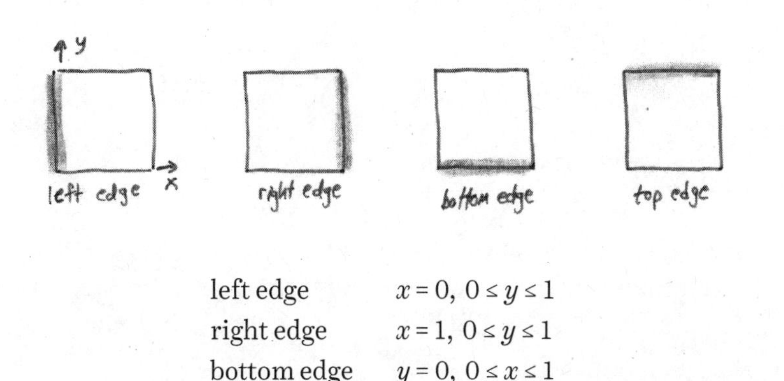

left edge	$x = 0,\ 0 \le y \le 1$
right edge	$x = 1,\ 0 \le y \le 1$
bottom edge	$y = 0,\ 0 \le x \le 1$
top edge	$y = 1,\ 0 \le x \le 1$

The unit cube C in xyz-space consists of all points (x, y, z) with $0 \le x \le 1$, $0 \le y \le 1$, and $0 \le z \le 1$. To find the pieces of the boundary of C, as with the square we set one of the coordinates to an extreme value, 0 or 1, and let the other two coordinates vary over the entire range [0,1]. So we see that the boundary of the cube is six faces, each of these faces is a square.

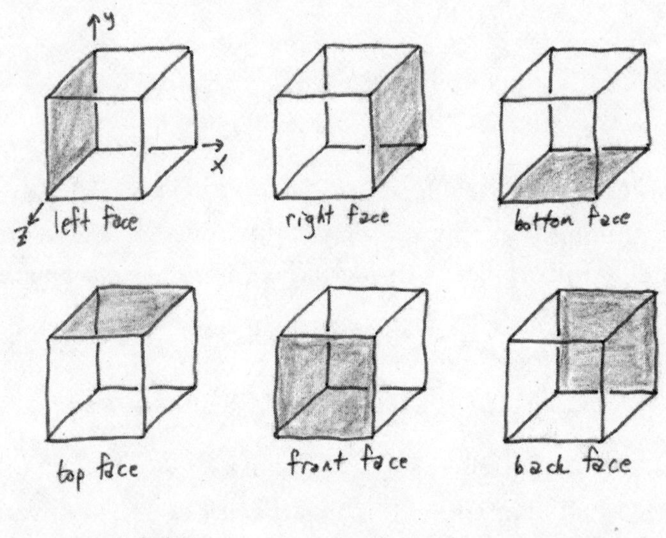

left face	$x = 0,\ 0 \le y \le 1,\ 0 \le z \le 1$
right face	$x = 1,\ 0 \le y \le 1,\ 0 \le z \le 1$
bottom face	$y = 0,\ 0 \le x \le 1,\ 0 \le z \le 1$
top face	$y = 1,\ 0 \le x \le 1,\ 0 \le z \le 1$
front face	$z = 1,\ 0 \le y \le 1,\ 0 \le x \le 1$
back face	$z = 0,\ 0 \le y \le 1,\ 0 \le x \le 1$

The unit hypercube H in $wxyz$-space consists of all points (w, x, y, z) with $0 \le w \le 1$, $0 \le x \le 1$, $0 \le y \le 1$, and $0 \le z \le 1$. To find the pieces of the boundary of the hypercube H, set one coordinate to an extreme value and let the other three range over [0,1]. For example, one boundary cube is

$$w = 0,\ 0 \le x \le 1,\ 0 \le y \le 1,\ 0 \le z \le 1$$

Each coordinate has two extreme values, so four coordinates means the boundary of a hypercube consists of eight cubes. Here are pictures of these eight cubes. In the top two images we shade in the two "obvious" cubes. Call the shaded cube on the left the *lower cube*, and the shaded cube on the right the *upper cube*.

The other six shaded cubes connect one of the faces of the lower cube to the corresponding face of the upper cube. For

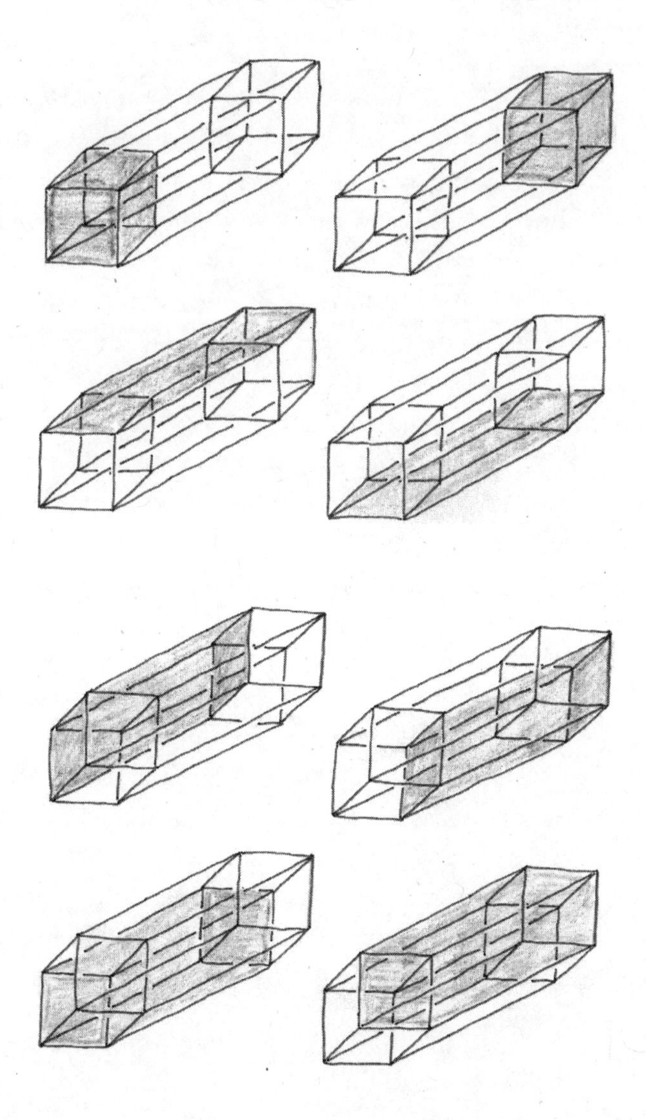

example, the left shaded cube in the second row connects the top face of the lower cube to the top face of the upper cube.

How many hypercubes make up the boundary of a five-dimensional cube?

WHY $\sqrt{2}$ IS IRRATIONAL

To show that the square root of 2 is not the ratio of two whole numbers, Mr. Griffith began: Suppose you could write $\sqrt{2}$ as a ratio of whole numbers, say $\sqrt{2} = a/b$, and let's take the fraction to be in lowest form (so we'd write 7/5 instead of 14/10, for example). Now square both sides, so $2 = a^2/b^2$, that is, $2b^2 = a^2$. Now, is a^2 even or odd? It's 2 times b^2, so a^2 is even. So is a even or odd? Well, the square of an even number is even, and the square of an odd number is odd, so a must be even. That means $a = 2c$ for some whole number c. Now go back to $2b^2 = a^2$. Do you see the problem? Umm, $2b^2 = a^2 = (2c)^2 = 4c^2$. Now cancel a 2 from both sides. What do you see? Oh, $b^2 = 2c^2$, so b^2 is even and b is even, and this is a problem because both a and b are even, but we'd taken the fraction a/b to be in lowest form. Hah! This is so cool.

SOME FINE PRINT ABOUT FRACTALS

When we say the gasket is the only shape left unchanged by the gasket rules, we must be a bit careful. The gasket isn't the only shape left unchanged. For example, if we apply the three gasket rules to the whole plane, we get the whole plane back again. What we can say is that the gasket is the only closed and bounded shape left unchanged by the three gasket rules.

A shape is *closed* if its complement is open, and a shape is *open* if every point in the shape is the center of a small disk that lies entirely inside the shape. For example, $\{(x, y) : x^2 + y^2 < 1\}$ is open and $\{(x, y) : x^2 + y^2 \leq 1\}$ is not open.

A shape is *bounded* if the whole shape can be encompassed in a large enough circle.

A BIT ABOUT FRACTAL DIMENSION

This section has *a lot* more math than the rest of the book. Here we'll sketch some of the geometry of dimensions introduced in chapter 5. We'll stick to simple geometry; all applications to the physical world are complicated by the inherent noisiness of nature. We introduced dimension by asking how many copies of a shape are formed if we double its width and height. A related approach can be generalized more easily. Rather than grow a shape, we'll keep it the same size and try to split it into smaller copies similar to the whole shape. We've already seen this decomposition for the Sierpinski gasket: it is made of three copies, each scaled by a factor of ½. Call the number of copies N and the scaling factor r. Then the fractal dimension d is given by

$$N = (1/r)^d$$

Why is it $1/r$? Because N is greater than 1, r is less than 1, and at least in these situations d is positive. To find d, take the log of both sides, use the fact from algebra that $\log((1/r)^d) = d\log(1/r)$, and solve for d:

$$d = \frac{\log(N)}{\log(1/r)}$$

The assumption that underlies this calculation is that the shape is self-similar, so this is called the *similarity dimension*. For the Sierpinski gasket the similarity dimension is

$$d = \frac{\log(3)}{\log(2)} \approx 1.58496$$

Suppose a shape is self-similar but the pieces are not scaled by the same factor. Maybe each of the N pieces has its own scaling factor, $r_1, \ldots, r_N$.

The similarity dimension formula has no place to put more than one scaling factor. But we can rewrite $N = (1/r)^d$ as

$$Nr^d = 1 \text{ that is, } \underbrace{r^d + \cdots + r^d}_{N \text{ terms}} = 1$$

Because we have a term for each scaling factor, this formulation of the similarity dimension equation can accommodate different factors:

$$r_1^d + \ldots + r_N^d = 1$$

This is called the *Moran equation.*

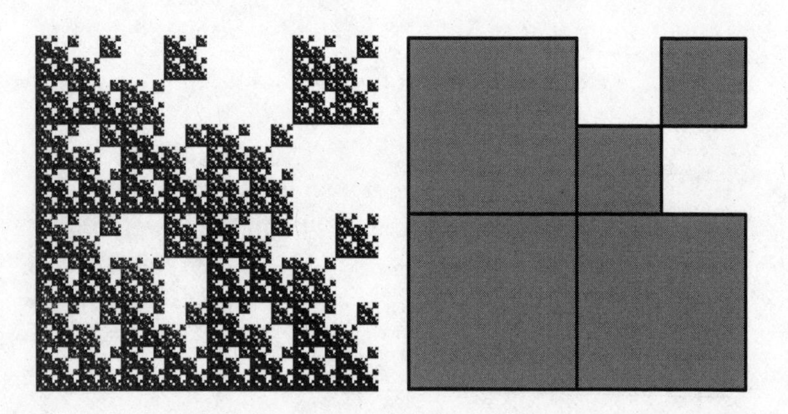

For example, here we see a fractal with several scaling factors. As indicated by the schematic diagram to the right, this fractal has

$$r_1 = r_2 = r_3 = 1/2$$

$$r_4 = r_5 = 1/4$$

and so the Moran equation becomes

$$3(1/2)^d + 2(1/4)^d = 1$$

Now you may think this must be solved numerically, because we can't solve for d by taking the log of both sides. But in this case there's another option, because $(1/4)^d = ((1/2)^2)^d = ((1/2)^d)^2$

Then we write $(1/2)^d = x$ and the Moran equation becomes the quadratic equation

$$3x + 2x^2 = 1$$

Apply the quadratic formula to find $x = (-3 \pm \sqrt{17})/4$. Because $x = (1/2)^d$ is positive, we take $x = (-3 + \sqrt{17})/4$. Finally, to find the value of d we take the log of both sides of

$$\left(\frac{1}{2}\right)^d = \frac{-3 + \sqrt{17}}{4}$$

and solve for d,

$$d = \frac{\log((-3 + \sqrt{17})/4)}{\log(1/2)} \approx 1.83251$$

We'll mention two extensions; there are many others. These results have several sources, all collected in chapter 6 of *Fractal Worlds: Grown, Built, and Imagined.*

First, we'll consider random fractals. By this we mean that instead of applying the same scaling factors at each iteration, each scaling factor can take one of several values with given probabilities. In this case the Moran equation is

$$\mathbb{E}(r_i^d) + \ldots + \mathbb{E}(r_N^d) = 1$$

where $\mathbb{E}(r_i^d)$ is the expected value, or average, of r_i^d. We call this the *randomized Moran equation.*

The picture below is a random fractal with $N = 4$ pieces, each with scaling factor $r = 1/2$ with probability $1/2$ and $r = 1/4$ with probability $1/2$. Then for each piece the expected value is $\mathbb{E}(r^d) = 1/2(1/2)^d + 1/2(1/4)^d$.

Again use $x = (1/2)^d$ so $x^2 = (1/4)^d$. Then the randomized Moran equation becomes the quadratic $2x + 2x^2 = 1$ and the dimension is

$$d = \frac{\log((-1 + \sqrt{3})/2)}{\log(1/2)} \approx 1.44998$$

But what is this number? Surely different sequences of choices of $1/2$ and $1/4$ will give different random fractals. The dimension we've calculated is the average value of the dimensions we'd obtain if we generated many fractals by this recipe.

Finally, let's return to the fractal on page 28 of chapter 1. This is generated by four transformations, all with scaling factor $r = \frac{1}{2}$, but with only some combinations allowed.

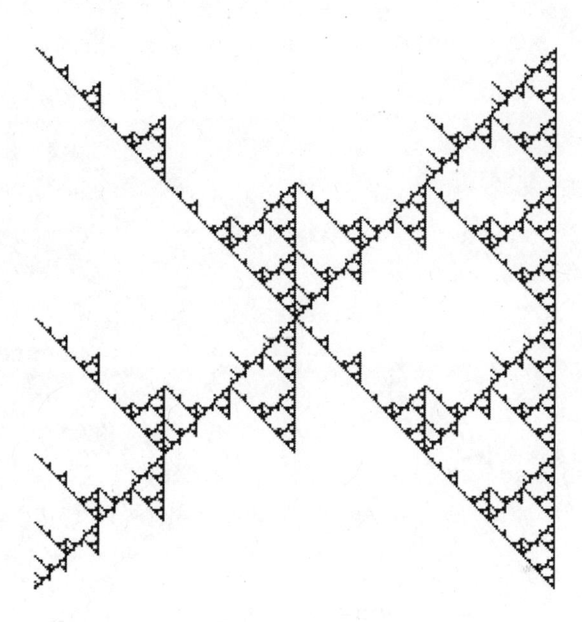

One way to express this is to label the quadrants of the fractal 1 (lower left), 2 (lower right), 3 (upper left), and 4 (upper right). We can encode which combinations are allowed and which are forbidden in a matrix. The number of the row represents a quadrant, and the number of the column represents a subquadrant of that quadrant. For example, the entry in the first row and second column corresponds to the lower right subquadrant within the lower left quadrant. A 0 in a matrix entry means that the corresponding subquadrant is unoccupied; a 1 means the subquadrant is occupied. Then the matrix that encodes the fractal shown above is

$$M = \begin{bmatrix} 1 & 0 & 1 & 1 \\ 0 & 1 & 1 & 1 \\ 0 & 1 & 1 & 0 \\ 1 & 1 & 0 & 1 \end{bmatrix}$$

Because all the scaling factors are equal, $r = ½$, what we could call the *memory Moran equation* is

$$(½)^d \rho[M] = 1$$

The factor $\rho[M]$ is called the spectral radius of M. It's the largest eigenvalue of the matrix M. Here we won't teach you to compute eigenvalues.

For that, see pretty much any linear algebra book, or Appendices A.83 and A.84 of *Fractal Worlds*. For this matrix M the eigenvalues are $1 \pm \sqrt{3}$, 1, and 1. The number of eigenvalues equals the number of rows (or columns) of the matrix, allowing for the fact that sometimes eigenvalues are repeated. The eigenvalue 1 here is an example. The spectral radius is $\rho[M] = 1 + \sqrt{3}$ and by solving the memory Moran equation for d we find

$$d = \frac{\log(\rho[M])}{\log(2)} \approx 1.44998.$$

The list of extensions of the Moran equation is more extensive. For instance, there is a version of the Moran equation if the contraction factor of a transformation varies with position. But we've seen enough for now.

A final comment about the Moran equation. In some of our examples we converted it to a quadratic equation. What do we do if the solutions of the quadratic equation are complex? This can't happen: the Moran equation always has a solution, and only one solution. See Appendix A.76 of *Fractal Worlds*.

Now for the bit of fine print about the measurement and dimension.

If we try to measure a shape in a dimension lower than the dimension of the shape, the answer we get is ∞; if we try to measure in a dimension higher than that of the shape, the answer we get is 0. The technicalities are pretty complicated, but an example will illustrate the idea. Suppose the shape is a filled-in unit square, certainly two-dimensional.

To measure its length, suppose we try to cover the square with infinitely thin thread. Any finite length of thread would leave many gaps, so we need an infinite length of thread to cover the square.

On the other hand, the square can fit inside a box with base a unit square and with height h for every $h > 0$. The volume of this box is $V = 1^2 h = h$. As we take smaller and smaller values of h, the volume of the box approaches 0, so the volume of the square is 0.

Here's where the fine print comes in: we'll consider only bounded shapes. An infinitely long but narrow strip in the plane has infinite area, but is two-dimensional. Getting an infinite value for a measurement can be the result of measuring in too low a dimension (this case interests us) or measuring an unbounded shape (this case doesn't interest us). So we'll stick with bounded shapes.

If all this is too abstract, we'll make it more concrete by thinking about the gasket. Suppose we build a gasket from a right isosceles triangle with base and altitude 1. Recall the area of a triangle is $\frac{1}{2} \times \text{base} \times \text{altitude}$, so this isosceles triangle has area $\frac{1}{2}$. We'll construct the gasket in another way, one that makes the area computation straightforward. Here's the construction: connect the midpoints of the sides of the filled-in triangle and remove the middle triangle. This leaves three filled-in triangles, so repeat the process for each of these and continue.

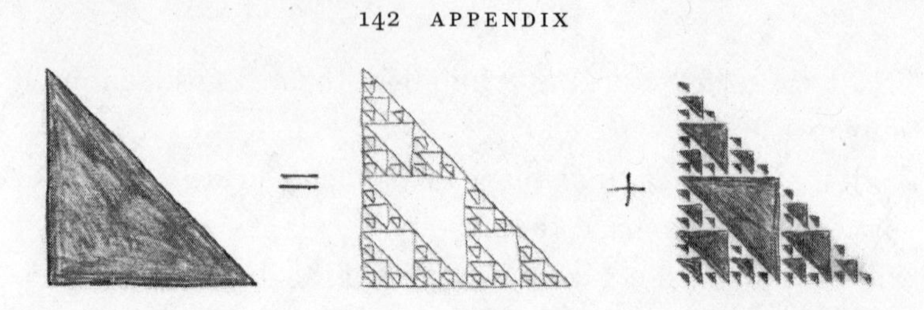

So we see that the original triangle can be decomposed into the gasket a collection of triangles. The largest of these removed triangles has has base and altitude ½, so has area ⅛. The next largest removed triangles, there are 3 of these, have base and altitude ¼, so has area 1/32. Continuing in this way, we can add up the areas of the removed triangles:

$$\frac{1}{8}+\frac{3}{32}+\frac{9}{128}+\cdots=\frac{1}{8}\left(1+\frac{3}{4}+\frac{3^2}{4^2}+\cdots\right)=\frac{1}{8}\,\frac{1}{1-3/4}=\frac{1}{2}$$

Here the next-to-the-last equality is an instance of summing a geometric series, for every ratio r with $|r| < 1$, the series $1 + r + r^2 + r^3 + \ldots$ sums to $1/(1-r)$. The areas of the removed triangles sum to ½, the area of the original triangle, so the area of the gasket is 0.

What would we mean by the length of the gasket? The sum of the perimeters of the triangles is a good first step. The perimeter may have more parts that we can't see, or it may not, but let's see what the perimeters tell us. The perimeter of the big right triangle is $1+1+\sqrt{2}=p$. The perimeter of the first removed triangle is $p/2$, the perimeters of the second removed triangles is $p/4$, and so on. Then the sum of these perimeters is

$$p+\frac{p}{2}+\frac{3p}{4}+\frac{9p}{8}+\cdots=p+\frac{p}{2}\left(1+\frac{3}{2}+\frac{3^2}{2^2}+\cdots\right)=\infty$$

The gasket is a bounded set in the plane with infinite length, so its dimension is > 1, and with zero area, so its dimension is < 2. Because the gasket dimension lies between consecutive integers, it cannot be an integer. This is how measurements can tell us about dimension.

Acknowledgments

Love is absolutely worth the risk of loss.

First, I must thank my aunt Ruth Frame. Endlessly curious, she was an adult who really listened to the ideas of little kids. She was the first person I lost. I was old enough to understand what had happened, but not old enough to have processed the stories adults tell about death. When she died, I felt the loss directly and unvarnished. Pure, paralyzing emotion.

But besides the pain of this loss, and more important to the way my life has unfolded, is how Ruthie amplified my curiosity, showed me that science is a path that a little kid can begin to follow. I wish she could have seen the adults that her nieces and nephews became. I wish I could have handed her a copy of *Chaos Under Control* or *Fractal Worlds* and told her they are results of the life she showed me how to live.[1]

My parents, Mary and Walter Frame; my siblings Steve Frame and Linda Riffle and their spouses Kim and David; my nephew Scott Lothes and his wife Maureen Muldoon; my cousin Matt Arrowood, his wife Susan, and their sons Zane and Will; and my wife Jean Maatta and I have woven complicated lives together.

Certainly grief is a part of all our lives, but our lives have ever so much more than grief. Even on the worst days, we have good moments. And on the best days, well . . .

Story space arose in many discussions with Amelia Urry before we began work on *Fractal Worlds*. A delightful independent study with Caroline Kanner and Caroline Sydney led to some variations of story space and many explorations of other aspects of geometry and literature. These were central to the analysis of chapter 4.

Specific ties between story space and grief benefited greatly from thoughtful and energetic discussions with Rich and Kayla Magliula. Rich is our veterinarian. He has taken, and continues to take, wonderful care of our family of cats.

My nephew Scott pointed out the story geometry in John McPhee's *Draft No. 4*.[2] Also, Scott and I have had many conversations about Murakami's books. Two thoughtful readers who approach a beloved author from different directions—I am a geometer and teacher, Scott is an author, photographer, and editor—can take quite dissimilar, and equally valid, ideas from a complex story.

Lara Santoro shared some of her ideas about grief. As a foreign correspondent who spent years covering the AIDS epidemic in Nairobi, her firsthand experiences with grief are substantial. Her novel *Mercy* is a fictionalized account of that part of her life; her novel *The Boy* describes another type of grief with complex moral implications.[3] In our correspondence and conversations, Lara described a method to cope with grief by zooming out. If you can become aware that you are aware, grief loses some of its sting. This idea redirected the analysis of chapter 5.

In correspondence with Andrea Sloan Pink we explored stories, loss, and how we perceive the world. Andrea and I began exchanging emails when I found that I am a character in one of her plays.[4] Andrea's mother died not long after my father died.

Our conversations about grief at the loss of a parent brought to the surface some of the variety of ways we grieve. This correspondence was enlightening, and at least to me, comforting.

My editor Joe Calamia drew the idea for this book from a conversation we had several years ago. This the third project that Joe and I have done together; working with him is an uncommon delight. This is my first book that is not an academic text. Joe has been a gentle, patient, and indefatigable guide to writing for a more general audience. Joel Score's careful reading of the text and insightful comments enabled me to see the flow of ideas as they will appear to others. His skill is another reason I enjoy working with the University of Chicago Press. I hope the three of us do more projects together.

My cousin Patti Reid read the manuscript, caught typos that I'd missed, and provided helpful comments about tone. She also gave me an image—us sitting at her kitchen table and trading stories—that has buoyed me through some difficult times. Yeah, family is important. In addition, Patti introduced me to her grandchildren Asterisk (Astrid) and BeetleBomb (Ira), whose curiosity and delight at the world provide still more evidence that love is absolutely worth the risk of loss.

Mike Donnally, the only friend from my childhood neighborhood full of kids with whom I'm still in touch, provided an important piece of information from those times that had evaporated from my memory. Thanks, Mike.

Paul Dunkle, another student in my high school and now a family member and friend, also provided some important information from those times. Thanks, Paul.

For this, and my three previous books, Andy Szymkowiak provided important technical assistance. Thanks, Andy.

In November 2012, I was on a panel with Laurie Santos. The panel, organized by a student group, began with a discussion of happiness, but the focus broadened under student questions. When an audience member asked about depression, Laurie let me answer even though I'm sure she could have said something more interesting. As my response unfolded, I saw a hint that geometry can shed light on emotion. The possibility of this book grew from that hint, which was in the back of my mind during that conversation with Joe Calamia. Thanks, Laurie, for letting me run with an idea that was half-baked then.

Thoughtful, detailed comments by two anonymous reviewers led to significant improvements in the both the architecture and the presentation of ideas. One reviewer pointed out the book *Mathematics and Humor* by John Allen Paulos, not because grief has any humorous aspect, but because in his chapter 5 Paulos develops a geometric theory of humor.[5] This helped me refine some aspects of my geometric theory of grief. Also, the supportive remarks of both reviewers gave me a way to cope with the worry that no one would be interested in this book. Reviewers can have a substantial impact on how an author goes forward with a project.

The patience, conversations, and kindness of my wife Jean Maatta have been essential for this book, and for every other book I've written. If I live another thirty years (not likely: I don't expect to make it to one hundred), I might understand how lucky I am that she agreed to marry me.

The loss of many cats showed me new dimensions of grief, helped direct some of my thoughts in this project. But then, too, a thousand hours of cats asleep on my lap have been awfully good company while I worked. I would not give this up in order to avoid the pain at their loss.

Grief is a part of every life. Can we use geometry in a way that lessens the pain grief causes? What do you think?

* * *

The photo of Ruth Frame at the start of the acknowledgments is from an old family album. Anyone who might have remembered the photographer is long dead. The photo of Asterisk and Beetle-Bomb was taken by John Kim.

The computer-generated images were produced by Mathematica code I wrote. Although you might think that the hand-drawn sketches were done by some ten-year-old relation, sadly they were drawn by me. Evidently, I did not use the last six decades to improve my drawing skills. I did spend the time learning math, physics, coding, and a bit of biology, but not, alas, art.

Notes

PROLOGUE

1. "Worse" because whatever you believe, denying the loss tarnishes the memory of the life. And because even little kids deserve honesty, honesty perhaps filtered and gentled, but honesty nonetheless. Tell kids it's okay to feel sad; don't ever tell them there's no reason to feel sad.

2. My dear friend Christine Waldron gave me a copy of Ethan Canin's book *A Doubter's Almanac* (New York: Random House, 2016). For years I've read and admired Canin's work. Christine brought me to this book sooner than I would have got there on my own—in fact, soon after the death of my father, which likely contributed to how I understand Canin's words.

3. John Archer, *The Nature of Grief* (New York: Taylor & Francis, 1999), and Barbara King, *How Animals Grieve* (Chicago: University of Chicago Press, 2014), are excellent sources for careful thoughts about grief. King's book is more narrative, more personal; Archer's is a more abstract treatise. In that sense, they present complementary approaches; both are useful. Chapter 2 of *The Nature of Grief* sketches a history of grief scholarship. Randolph Nesse, "An Evolutionary Framework for Understanding Grief," in *Spousal Bereavement in Late Life*, ed. D. Carr, R. Nesse, and C. Wortman, 195–226 (New York: Springer, 2005), gives a clear exposition of the evolutionary basis of grief, a facet of his and George Williams's groundbreaking work on Darwinian medicine; see Nesse and Williams, *Why We Get Sick: The New Science of Darwinian Medicine* (New York: Random House, 1994).

4. Alexander Shand, *The Foundations of Character* (London: Macmillan, 1914), is the first systematic study of grief.

5. In chapter 3 of *The Nature of Grief*, Archer presents studies of grief viewed through art.

6. These books by Jean-Paul Sartre gave me my first, and still my clearest, picture of fiction as be the cleanest route to deep truths. Sartre gives a careful

philosophical analysis in *Being and Nothingness: A Phenomenological Essay on Ontology* (New York: Washington Square, 1966). From its subtitle on, the book is not particularly welcoming. In the novels comprising *The Roads to Freedom—The Age of Reason*, *The Reprieve*, and *Iron in the Soul* (all New York: Penguin, 1963)—the narratives of the characters lead to similar understandings. Stories resonate.

7. Music adds vocalization, emotion diamond-etched in the catch of a voice, underscored by pace. To which may be further added lone keyboard meditation, ensemble instruments, or the complex web of a crowd sounding over one another. The poetry of text can be expanded by, or replaced by, acoustic flights. Music thus lets us feel the richness of a life more directly and on more levels than may be apparent in a text. A few examples: Natalie Merchant, "My Skin," on *Ophelia* (Elektra, 1998), and "Beloved Wife," on *Tiger Lily* (Elektra, 1995); Loreena McKennitt, "Dante's Prayer," on *The Book of Secrets* (Quinlan Road, 1997); Philip Glass, "Knee 5," on *Einstein on the Beach* (Elektra, 1993). I could think of dozens, maybe hundreds, more. So could you. I wonder how much our lists would overlap?

8. Ang Lee's beautiful film *Crouching Tiger, Hidden Dragon* (Columbia Pictures, 2000) concludes with the breathtaking (literally—at least, it took my breath away, and I was not the only person in the theater who reacted that way if the stifled sobs I heard were an indicator) "Farewell" performed by Yo-Yo Ma.

9. Sia's "Breathe Me" is on *Color the Small One* (Astralwerks, 2006).

10. Parents should not outlive their children. Long before their parents died, my mother's younger brother Bill died of lung cancer, the result of two packs a day for forty years. Their grief was even greater because they tried to be brave. But who is that brave? Turgenev's portrait in *Fathers and Sons* (New York: Penguin, 1965) of the grief of Yevgény's parents is simple, straightforward, and moving. Really moving.

11. The earliest well-known empirical study of grief is Erich Lindemann, "Symptomatology and Management of Acute Grief," *American Journal of Psychiatry* 101 (1944): 141–48. Lindemann extended his analysis to include "anticipatory grieving," by which he meant an emotional response when one expects a loved one to die. This is the one exception to a strict requirement of the irreversibility of grief.

CHAPTER ONE

1. Grünbaum and Shephard, *Tilings and Patterns* (New York: Freeman, 1987).

2. The proof that there are exactly seventeen wallpaper groups was presented in Evgraf Fedorov, "Symmetry in the Plane," *Proceedings of the Imperial St. Petersburg Minerological Society* 28 (1891): 345–90. So why do we need the

proof? Because without the proof there might be an eighteenth wallpaper group—a tiling pattern that offers a new form of artistry—that lurks somewhere in the folds of geometry and so far has gone unnoticed.

3. The Cattedrale di Santa Maria Annunziata in Anagni, Italy, was completed in 1104. The interior tilework, including the Sierpinski gasket illustrated in the text, was added in the following century. Étienne Guyon and H. Eugene Stanley, *Fractal Forms* (Haarlem: Elsevier, 1991), called attention to the fractal aspects of the mosaic. See the photograph at https://commons.wikimedia.org/wiki/File:Anagni_katedrala_04.JPG.

4. You can see a *much* better rendition of the painting *Visage of War* at Wikipedia, https://en.wikipedia.org/wiki/The Face of War, and on page 97 of Robert Descharnes, *Dalí* (New York: Abrams, 1985), which includes a copy of a preliminary study of the painting.

5. The video "Linear Perspective: Brunelleschi's Experiment," https://www.youtube.com/watch?v=bkNMM8uiMww, demonstrates the mirror experiment.

6. Banchoff's book, *Beyond the Third Dimension: Geometry, Computer Graphics, and Higher Dimensions* (New York: Freeman, 1990), might have been titled "Thirteen Ways of Looking at a Hypercube," with apologies to Wallace Stevens and Henry Louis Gates Jr.

7. Sources of pictures of Dalí's *Crucifixion (Corpus Hypercubus)* are Wikipedia, https://en.wikipedia.org/wiki/Crucifixion_(Corpus_Hypercubus); Banchoff, *Beyond the Third Dimension*, 105; and Descharnes, *Dalí*, 119. On page 110 of Banchoff's book we see a photograph of Banchoff talking with Dalí.

8. H. S. M. Coxeter, *Non-Euclidean Geometry*, 5th ed. (Toronto: University of Toronto Press, 1965), and Marvin Greenberg, *Euclidean and Non-Euclidean Geometries: Development and History*, 4th ed. (New York: Freeman, 2007), are good references for non-Euclidean geometry. The Wikipedia page https://en.wikipedia.org/wiki/Non-Euclidean_geometry also is a reasonable place to start. The webpage https://brewminate.com/escher-and-coxeter-a-mathematical-conversation/ recounts the correspondence between Escher and Coxeter.

9. Copies of Escher's *Circle Limit III* can be found at Wikipedia, https://en.wikipedia.org/wiki/Circle Limit III, and in the book *M. C. Escher: 29 Master Prints* (New York: Abrams, 1983).

10. Sean Carroll, *Something Deeply Hidden: Quantum Worlds and the Emergence of Spacetime* (New York: Dutton, 2019).

11. Some details, and more examples, of fractals generated with memory are given in section 2.5 of Michael Frame and Amelia Urry, *Fractal Worlds: Grown, Built, and Imagined* (New Haven, CT: Yale University Press, 2016).

12. This result, known as *Gödel's incompleteness theorem*, was such a sur-

prise, and the idea underlying the proof so brilliant, that Kurt Gödel and Albert Einstein often took walks together on the grounds of the Institute for Advanced Study at Princeton. Einstein remarked that he often went to his office "just to have the privilege of walking home with Kurt Gödel." The story of their friendship is told beautifully in Jim Holt, *When Einstein Walked with Gödel: Excursions to the Edge of Thought* (New York: Farrar, Straus, and Giroux, 2018); the Einstein quote is from page 4. Ernest Nagel and James Newman's book *Gödel's Proof* (New York: New York University Press, 1958) gives a clear, compact explanation of the proof of Gödel's incompleteness theorem. Douglas Hofstadter's amazing book *Gödel, Escher, Bach: An Eternal Golden Braid* (New York: Basic Books, 1979) gives a significantly less compact, and wonderfully entertaining, explanation of Gödel's incompleteness theorem, each chapter introduced by an inventive fable about Achilles, the tortoise, and their friends.

13. The proof that these three classical problems of Greek geometry cannot be solved uses a branch of math called Galois theory, clearly presented in Ian Stewart, *Galois Theory*, 2nd ed. (London: Chapman and Hall, 1973). By clever arguments these three geometry problems are converted into algebraic constructions which are shown to be impossible. Perhaps surprisingly, some of these problems can be solved in non-Euclidean geometry.

14. The sequence of pictures turning a sketch of a cat into a gasket is shown in the figure on page 86. You'll know it when you see it.

15. Martin Gardner, *aha! Insight* (New York: Freeman, 1978).

16. Jorge Luis Borges, *Labyrinths: Selected Stories and Other Writings* (New York: New Directions, 1964), is, I think, the best introduction to the energetic imagination of Borges.

17. For example, Borges wrote a review of Edward Kasner and James Newman, *Mathematics and the Imagination* (New York: Simon & Schuster, 1940). The review is reprinted in Jorge Luis Borges, *Selected Non-Fictions*, ed. E. Weinberger, 249–50 (New York: Penguin, 2000).

18. Borges's story "The Library of Babel," the sixth story of *Labyrinths*, and his essay "Avatars of the Tortoise," the seventh essay of *Labyrinths*, are good examples of paradoxes and puzzles.

19. Borges describes some of the arithmetic of infinities in the first few pages of his essay "The Doctrine of Cycles," in *Selected Non-Fictions*, 115–22.

20. Borges presents some interesting variations in "Tlön, Uqbar, Orbis Tertius," the first story of his collection *Labyrinths*.

21. Not as clear a circle as José Saramago's *Death with Interruptions* (Orlando: Harcourt, 2005), but a circle nonetheless. Saramago's short novel presents a beautiful, self-contained circular story geometry. We analyze part of this story in more detail in chapter 3.

22. Borges, “Circular Time,” in *Selected Non-Fictions*, 225–28.

23. How big is $10^{10^{118}}$? The number 10^{118} is 1 with 118 zeroes after it, and so $10^{10^{118}}$ is 1 with 10^{118} zeroes after it. Is this all that big? The number of particles in the observable universe is about 10^{80}, so 10^{118}, the number of *zeros* in $10^{10^{118}}$, is the number of particles in 10^{38} copies of the observable universe.

24. Now for some cosmology. See Max Tegmark’s article “Parallel Universes,” *Scientific American* 288 (May 2003): 40–51, and his book about the mathematization of cosmology, *Our Mathematical Universe: My Quest for the Ultimate Nature of Reality* (New York: Knopf, 2014). The basic ideas are these: “There exists an external physical reality completely independent of us humans,” and “Our external physical reality is a mathematical structure.” This is a very interesting book.

25. The original calculations on how the Big Bang model could account for the observed cosmic abundances of the lightest elements were presented in Ralph Alpher, Hans Bethe, and George Gamow, “The Origin of Chemical Elements,” *Physical Review* 73 (1948): 803–4. (Gamow, known as a jokester, added Bethe’s name as an author so the paper would be referred to as the Alpher-Bethe-Gamow, α-β-γ, paper. Really.) Much more detail, along with updated astronomical evidence, is in Alpher’s chapter “Origins of Primordial Nucleosynthesis and Prediction of Cosmic Background Radiation,” in *Encyclopedia of Cosmology: Historical, Philosophical, and Scientific Foundations of Modern Cosmology*, ed. N. Hetherington, 453–75 (New York: Garland, 1993).

26. Sean Carroll, *From Eternity to Here: The Quest for the Ultimate Theory of Time* (New York: Dutton, 2010), is a crystal-clear meditation on the source of the directionality of time. For Boltzmann’s approach to the low-entropy problem of the early universe, see the figures on pages 213 and 216, and the text associated with those figures. Carroll describes some objections to Boltzmann’s approach in chapter 10, “Recurrent Nightmares,” and beautifully explains the baby universes argument in chapter 15, “The Past through Tomorrow.”

27. In his “Personal Note,” in S. Lloyd, *Programming the Universe: A Quantum Computer Scientist Takes on the Cosmos* (New York: Random House, 2006), 213–16, Seth Lloyd reports that while on a hike with his friend Heinz Pagels, Pagels slipped, fell several hundred feet down a sheer cliff, and died. Lloyd thought about the many worlds model, that in some, perhaps many, parallel universes his friend did not fall. This was no comfort for Lloyd. Andrea Sloan Pink reminded me of this passage, and reported that she and her children agree with Lloyd. Rather, Lloyd said, “Consolation has gradually come from information—from bits both real and imagined.” The person is gone, but their ideas, and our memories of their behaviors, remain for a while. Lloyd also reports (pages 101–2) that he met Borges at Cambridge and asked the writer if

the many worlds model was the inspiration for "The Garden of Forking Paths." Borges's replied, "No," and added that he wasn't surprised that the laws of physics reflect ideas in literature, because physicists do read literature.

CHAPTER TWO

1. Poincaré's formulation of chaos that arose in his efforts to determine the stability of the solar system is translated in Henri Poincaré, *New Methods in Celestial Mechanics*, ed. D. Goroff (American Institute of Physics, 1993). George Birkhoff and Jacques Hadamard discovered chaos in motion on saddle-shaped surfaces: G. Birkhoff, "Quelques théorèms sur le mouvement des systèmes dynamiques," *Bulletin de la Société Mathématique de France* 40 (1912): 305–23; J. Hadamard, "Les surfaces à courbures opposées et leur lignes geodesics," *Journal de Mathématiques* 4 (1898): 27–73. Lucy Cartwright and John Littlewood discovered chaos in the dynamics of radar circuits: L. Cartwright and J. Littlewood, "On Non-Linear Differential Equations of the Second Order I: The Equation $y'' + k(1 - y^2) + y = b\lambda k \cos(\lambda t + a)$, k large," *Journal of the London Mathematical Society* s1-20 (1942): 180–89. Edward Lorenz discovered chaos in early computer simulations of atmospheric convection models: E. Lorenz, "Deterministic Non-Periodic Flows," *Journal of the Atmospheric Sciences* 20 (1963): 130–41. Robert May discovered chaos in simple models of population dynamics with limited resources: R. May, "Simple Mathematical Models with Very Complicated Dynamics," *Nature* 261 (1976): 459–67. May's paper catalyzed a deluge of work in experimental mathematics. James Gleick's *Chaos: Making a New Science* (New York: Viking, 1987), a popular account of the discovery of chaos, was a *New York Times* bestseller.

2. C. S. Lewis gives an account of his grief at the loss of his wife in *A Grief Observed* (New York: Harper Collins, 1961).

3. Joan Didion, *The Year of Magical Thinking* (New York: Random House, 2005) and *Blue Nights* (New York: Random House, 2011). With the deaths of her husband and her daughter within two years, Didion has suffered a lot of grief in a short time. Then, almost two years after her daughter died, Didion developed shingles. Enough misery for one person.

4. Peter Heller, *The Dog Stars* (New York: Knopf, 2012).

5. So when Erich Lindemann, "Symptomatology and Management of Acute Grief," *American Journal of Psychiatry* 101 (1944): 141–48, and Colin Parkes, "Anticipatory Grief," *British Journal of Psychiatry* 138 (1981): 183, discuss anticipatory grief, I'll give them a pass.

6. John Archer, *The Nature of Grief* (New York: Taylor & Francis, 1999). In chapter 6, Archer describes efforts to find low-dimensional models of grief. His comment about the phase view of grief is on page 100.

7. John Bowlby, *Attachment and Loss*, volume 3, *Loss: Sadness and Depression* (London: Hogarth, 1980).

8. Colin Parkes, *Bereavement: Studies of Grief in Adult Life* (London: Tavistock, 1972).

9. Alexander Shand, *The Foundations of Character* (London: Macmillan, 1914).

10. Some arguments opposing the stages or phases view of grief are presented in Archer, *Nature of Grief*, 28, 29, 100.

11. Some opposition to the grief work hypothesis is presented in Archer, *Nature of Grief*, 122, 251, and in W. Stroebe, M. Stroebe, and H. Schut, "Does 'Grief Work' Work?" *Bereavement Care* 22 (2009): 3–5.

12. Stroebe and Schut have written a *lot* of papers on grief. The dual process model was presented three conference talks: M. Stroebe and H. Schut, "Differential Patterns of Coping with Bereavement between Widows and Widowers," British Psychological Society Social Psychology Section Conference, Jesus College, Oxford, 22–24 September 1993; M. Stroebe and H. Schut, "The Dual Process Model of Coping with Bereavement," Fourth International Conference on Grief and Bereavement in Contemporary Society, Stockholm, 12–16 June 1994; M. Stroebe and H. Schut, "The Dual Process Model of Coping with Loss," International Work Group on Death, Dying and Bereavement, St. Catherine's College, Oxford, 26–29 June 1995. An update is given in M. Stroebe and H. Schut, "The Dual Process Model of Coping with Grief: A Decade On," *Omega* 61 (2010): 237–89.

13. Some details on kin selection are presented in Sonya Bahar, *The Essential Tension: Competition, Cooperation, and Multilevel Selection in Evolution* (New York: Springer, 2018); William D. Hamilton, "The Genetic Evolution of Social Behavior, I and II," *Journal of Theoretical Biology* 7 (1964): 1–52; Oren Harman, *The Price of Altruism: George Price and the Search for the Origins of Kindness* (New York: Norton, 2010); Martin Nowak and Roger Highfield, *Supercooperators: Altruism, Evolution, and Why We Need Each Other to Succeed* (New York: Simon & Schuster, 2011); Richard Prum, *The Evolution of Beauty: How Darwin's Forgotten Theory of Mate Choice Shapes the Animal World—and Us* (New York: Doubleday, 2017); and Prum's TEDxYale talk, https://www.youtube.com/watch?v=128-i8ulC7o.

14. Barbara King, *How Animals Grieve* (Chicago: University of Chicago Press, 2014).

15. Evidence for episodic and autobiographical memory in animals is presented in Gema Martin-Ordas, Dorthe Bernsten, and Josep Call, "Memory for Distant Past Events in Chimpanzees and Orangutans," *Current Biology* 23 (2013): 1438–41.

16. See, for example, King, *How Animals Grieve*, 85.

17. King, *How Animals Grieve*, chap. 6, describes female monkeys carrying dead infants for days.

18. Helen Macdonald, *H Is for Hawk* (New York: Grove Press, 2014).

19. An evolutionary basis for grief is presented in Randolph Nesse, "An Evolutionary Framework for Understanding Grief," in *Spousal Bereavement in Late Life*, ed. D. Carr, R. Nesse, and C. Wortman, 195–226 (New York: Springer, 2005).

20. Medicine understood through the lens of Darwinian evolution is explored through many examples in Randolph Nesse and George Williams, *Why We Get Sick: The New Science of Darwinian Medicine* (New York: Random House, 1994).

CHAPTER THREE

1. The quotation is from Barbara King, *How Animals Grieve* (Chicago: University of Chicago Press, 2014), 14.

2. More recently, Hume, Kant, Schopenhauer, and especially Santayana theorized about beauty. Santayana's theory of aesthetics is presented in a book based on his Harvard lectures from 1892 to 1895, *The Sense of Beauty: Being the Outlines of Aesthetic Theory* (New York: Scribner, 1896). John Timmerman, *Robert Frost: The Ethics of Ambiguity* (Lewisburg, PA: Bucknell University Press, 2002), 174, relates that Santayana wrote the book, which he called a "wretched potboiler," to secure tenure at Harvard.

3. Berlyne's book on aesthetics, in which he presents his "novelty and familiarity" thesis, and his paper on curiosity are *Aesthetics and Psychobiology* (New York: Appleton-Century-Crofts, 1971) and "A Theory of Human Curiosity," *British Journal of Psychology* 45 (1954): 180–91.

4. Santayana presents his argument for the balance of purity and variety in section 16 of *The Sense of Beauty*.

5. Dutton's Darwinian theory of beauty is presented in *The Art Instinct: Beauty, Pleasure, and Human Evolution* (New York: Bloomsbury, 2009) and his TED talk, https://www.ted.com/talks/denis_dutton_a_darwinian_theory_of_beauty.

6. Ang Lee, director, *Crouching Tiger, Hidden Dragon* (Columbia Pictures, 2000).

7. José Saramago, *Death with Interruptions* (Orlando: Harcourt, 2005).

8. Our examples of art that can be appreciated independently of one's culture are Mbuti bark-cloth paintings, seen in figure 4.3 of Ron Eglash, *African Fractals: Modern Computing and Indigenous Design* (New Brunswick, NJ: Rutgers University Press, 1999); Inuit animal sculptures, seen in figure 133 of

Bernadette Driscoll, *Uumajut: Animal Imagery in Inuit Art* (Winnipeg, MB: Winnipeg Art Gallery, 1985); and the Lascaux cave paintings and mosque in Cordoba, Spain, seen on pages 74–77 and 289–99 of H. W. Janson, *History of Art*, 4th ed. (New York: Abrams, 1991).

9. Possibly this book: Douglas Hall, *Klee* (Oxford: Phaidon, 1977).

10. Charles Darwin, *On the Origin of Species by Means of Natural Selection, or the Preservation of Favoured Races in the Struggle for Life* (London: John Murray, 1859) and *The Descent of Man, and Selection in Relation to Sex* (London: John Murray, 1871).

11. Richard Prum's analysis of aesthetic selection is beautifully described in *The Evolution of Beauty: How Darwin's Forgotten Theory of Mate Choice Shapes the Animal World—and Us* (New York: Doubleday, 2017), and his TEDxYale talk, https://www.youtube.com/watch?v=128-i8ulC7o.

12. Ronald Fisher, "The Evolution of Sexual Preference," *Eugenics Review* 7 (1915): 184–91.

13. Amotz Zahavi, "Mate Selection: A Selection for a Handicap," *Journal of Theoretical Biology* 53 (1975): 205–14.

14. Mark Kirkpatrick, "The Handicap Mechanism of Sexual Selection Does Not Work," *American Naturalist* 127 (1986): 222–40; Alan Grafen, "Sexual Selection Unhandicapped by the Fisher Process," *Journal of Theoretical Biology* 144 (1990): 473–516.

15. A function $y = f(x)$ is linear if the change in y is proportional to the change in x, and nonlinear if the change in y is not proportional to the change in x. For example, $y = 5x$ is linear because for any change in x, the change in y will be proportional—in this case, five times greater. On the other hand, $y = x^2$ is nonlinear because if we multiply x by, say, 2, we multiply y by 4, while if we multiply x by 3, we multiply y by 9; the change in y is not proportional to the change in x.

16. The quote is from Prum, *Evolution of Beauty*, 186, 188.

17. Sewall Wright's notion of fitness landscape is described in his "Evolution in Mendelian populations," *Genetics* 16 (1931): 97–159, and "The Role of Mutation, Inbreeding, Crossbreeding, and Selection in Evolution," *Proceedings of the Sixth International Congress of Genetics* 1 (1932): 356–66.

18. Katherine Johnson is the subject of the wonderful film *Hidden Figures* (20th Century Fox, 2016).

19. Gödel numbering is described in E. Nagel and J. Newman, *Gödel's Proof* (New York: New York University Press, 1958), and in D. Hofstadter, *Gödel, Escher, Bach: An Eternal Golden Braid* (New York: Basic Books, 1979).

20. The quote is from Carl Sagan, *Cosmos* (New York: Random House, 1980), 4.

21. There's a bit of fine print associated with the statement that the gasket is the only shape left unchanged by the simultaneous application of the three gasket rules. The fine print is spelled out in the appendix.

22. The Mandelbrot set is described and illustrated in B. Mandelbrot, *The Fractal Geometry of Nature* (New York: Freeman, 1983), chap. 19. The general public first saw the Mandelbrot set in A. K. Dewdney, "Computer Recreations: Exploring the Mandelbrot Set," *Scientific American* 253 (August 1985): 16–21, 24, the cover article for the issue. The discovery of the Mandelbrot set is recounted in chapter 25 of Benoit's memoirs, B. Mandelbrot, *The Fractalist: Memoir of a Scientific Maverick* (New York: Random House, 2012).

23. The paper Dave Peak and I wrote to present our work with Henry Hurwitz is H. Hurwitz, M. Frame, and D. Peak, "Scaling Symmetries in Nonlinear Dynamics: A View from Parameter Space," *Physica D* 81 (1995): 23–31.

CHAPTER FOUR

1. There's some confusion about comparing the branching of the garden in Borge's "The Garden of Forking Paths," in *Labyrinths: Selected Stories and Other Writings*, to the branching of Hugh Everett's many worlds interpretation of quantum mechanics, in B. DeWitt and N. Graham, eds., *The Many-Worlds Interpretation of Quantum Mechanics* (Princeton, NJ: Princeton University Press, 1973). As very well explained by Sean Carroll in *Something Deeply Hidden* (New York: Dutton, 2019): macroscopic choices do not divide the universe into two branches; only measurements of quantum states do that, if many worlds does model the universe.

2. Carlo Rovelli's wonderful books on the nature of time and of reality from the perspective of modern quantum physics and relativity are *Reality Is Not What It Seems: The Journey to Quantum Gravity* (Penguin Random House, 2017) and *The Order of Time* (Penguin Random House, 2018).

3. Kurt Vonnegut's "Here Is a Lesson in Creative Writing" is chapter 3 of *A Man without a Country* (New York: Random House, 2007).

4. John McPhee draws parallels between geographical and narrative shapes in the "Structure" chapter of *Draft No. 4: On the Writing Process* (New York: Farrar, Straus and Giroux, 2017).

5. Bill Bryson, *A Walk in the Woods* (New York: Broadway, 1999).

6. Hal Ashby's film *Being There* (United Artists, 1979) is based on Jerzy Kosínski's novel *Being There* (Toronto: Bantam, 1970).

7. Leslie Jamison, *The Empathy Exams* (Minneapolis: Graywolf Press, 2014).

8. When you look at the figure you may think there's no way that the large jump in the *Scruffy play-t* plane could project to the small jump in the shaded plane. But remember, the path in the shaded plane is not a projection of the

path in the *Scruffy play-t* plane. Both are projections of a path in a higher-dimensional space.

CHAPTER FIVE

1. Saramago wrote about the fractal cemetery architecture of his novel *All the Names* (San Diego: Harcourt, 1997) in the March 31 entry in J. Saramago, *The Notebook* (London: Verso, 2010).

2. The first meditation, or manifesto, on fractals is B. Mandelbrot, *The Fractal Geometry of Nature* (New York: Freeman, 1983). Since then we've seen a vast collection of books: for little kids, S. Campbell and R. Campbell, *Mysterious Patterns: Finding Fractals in Nature* (Honesdale, PA: Boyds Mills, 2014); for general readers, K. Falconer, *Fractals: A Very Short Introduction* (Oxford: Oxford University Press, 2013); for teachers, M. Frame and B. Mandelbrot, *Fractals, Graphics, and Mathematics Education* (Washington, DC: Mathematical Association of America, 2002); for undergraduates, K. Falconer, *Fractal Geometry: Mathematical Foundations and Applications*, 3rd ed. (Chichester: Wiley, 2014), Michael Frame and Amelia Urry, *Fractal Worlds: Grown, Built, and Imagined* (New Haven, CT: Yale University Press, 2016), D. Peak and M. Frame, *Chaos under Control: The Art and Science of Complexity* (New York: Freeman, 1994), H.-O. Peitgen, H. Jürgens, and D. Saupe, *Chaos and Fractals: New Frontiers in Science*, 2nd ed. (New York: Springer, 2004), and Y. Pesin and V. Climenhaga, *Lectures on Fractal Geometry and Dynamical Systems* (Providence, RI: American Mathematical Society, 2009); for graduate students, K. Falconer, *Techniques in Fractal Geometry* (Chichester: Wiley, 1997); and scores, maybe hundreds, of volumes of conference proceedings.

3. We'll say a bit about how to calculate dimensions in the appendix.

4. Some details of the length and area computation for the gasket, and of the relation between dimension and measure, are presented in the appendix.

5. A simple speculation about life in fractional dimensions is presented in section 6.7 of Frame and Urry, *Fractal Worlds*.

6. Helen Macdonald, *H Is for Hawk* (New York: Grove Press, 2014).

7. Ann Pancake's novel *Strange as This Weather Has Been* (Berkeley, CA: Counterpoint, 2007) recounts the effects, mostly on members of one family, of environmental disasters caused by surface mining of coal in southern West Virginia. Near where I grew up; near when I grew up. This is not a preachy, boring tome. It is stories about complicated, flawed people who suffered through the consequences of ridiculous greed and stupidity.

CHAPTER SIX

1. Judea Pearl, *Causality: Models, Reasoning, and Inference*, 2nd ed. (Cambridge: Cambridge University Press, 2009); Judea Pearl and Dana Mackenzie,

The Book of Why: The New Science of Cause and Effect (New York: Basic Books, 2018).

2. *Roger Ebert's Journal*, 14 January 2009, https://www.rogerebert.com/rogers-journal/i-feel-good-i-knew-that-iwould.

3. From the dedication of Pearl, *Causality*.

ACKNOWLEDGMENTS

1. Dave Peak and I wrote *Chaos under Control: The Art and Science of Complexity* (New York: Freeman, 1994), as a text for our course on fractals and chaos for non-science students at Union College. About twenty years later Amelia Urry and I wrote *Fractal Worlds: Grown, Built, and Imagined* (New Haven, CT: Yale University Press, 2016), for the corresponding course at Yale. In the years between these volumes, the field itself grew, as did my understanding of it. Ironically, *Fractal Worlds* was published shortly after I retired from Yale, so I've not used it as a text.

2. John McPhee is one of the most remarkable nonfiction writers around. I've heard McPhee described as having only one trick, and that trick is to make you deeply interested in the subject of his current book, whatever that subject may be. That's quite a trick. J. McPhee, *Draft No. 4: On the Writing Process* (New York: Farrar, Straus and Giroux, 2017), describes his writing process and includes some geometry.

3. Lara Santoro's novels are about grief and loss. The first, *Mercy* (New York: Other Press, 2007), is set in AIDS-ravaged Africa; the second, *The Boy* (New York: Little, Brown, 2013), in the American southwest—and here the grief is more internal. Both are powerful, brilliant, and direct. I impatiently await her next novel.

4. Imagine my surprise to find that I am a character in Andrea Sloan Pink's play "Fractaland"; in *The Best American Short Plays, 2013–2014*, ed. W. W. Demastes, 249–63 (Milwaukee: Applause Theatre & Cinema Books, 2015). She so accurately portrayed Benoit and me that I wondered if she had been a student of mine; in fact, we've not met, but we did begin an exchange of emails, which still continues. That we both had lost a parent at about the same time made discussions of grief a natural part of our correspondence.

5. John Allen Paulos's geometrical model of humor is presented in *Mathematics and Humor* (Chicago: University of Chicago Press, 1980).

Index